Engineering Design:
A Project-Based Introduction

Engineering Design:
A Project-Based Introduction

Clive L. Dym

Harvey Mudd College

Patrick Little

Harvey Mudd College

John Wiley & Sons, Inc.

New York / Chichester / Brisbane / Weinheim / Singapore / Toronto

ACQUISITIONS EDITOR	Joseph Hayton
MARKETING MANAGER	Katherine Hepburn
SENIOR PRODUCTION EDITOR	Robin Factor
ILLUSTRATION EDITOR	Sigmund Malinowski
PHOTO EDITOR	Jill Hilycord
COVER DESIGNER	Miriam Dym
ELECTRONIC ILLUSTRATIONS	Radiant Illustration & Design

Cover Credits: Citicorp building and John Hancock building: Clive Dym. Ladders and gears: Miriam Dym. Beverage conveyor belt: Dennis O'Clair/Tony Stone Images.

This book was set in New Times Roman by Argosy and printed and bound by Courier/Stoughton. The cover was printed by Phoenix Color.

This book is printed on acid-free paper. ∞

Library of Congress Cataloging in Publication Data:
Dym, Clive L.
    Engineering design : a project-based introduction / Clive L. Dym and Patrick Little.
        p.   cm.
    ISBN 0–471–28296–0 (pbk. : alk. paper)
    1. Engineering design. I. Little, Patrick. II. Title.
TA174.D958   1999
620'.0042–dc21   99–26896
                                                          CIP

Printed in the United States of America

10 9 8 7 6 5 4 3

To

Joan Dym
whose love and support are distinctly nonquantifiable

and to

Charlie Hatch
a teacher's teacher

Foreword

To design is to imagine and specify things that don't exist, usually with the aim of bringing them into the world. The "things" may be tangible—machines and buildings and bridges; they may be procedures—the plans for a marketing scheme or an organization or a manufacturing process, or for solving a scientific research problem by experiment; they may be works of art—paintings or music or sculpture. Virtually every professional activity has a large component of design, although usually combined with the tasks of bringing the designed things into the real world.

Design has been regarded as an art rather than a science. A science proceeds by laws, which can sometimes even be written in mathematical form. It tells you how things must be, what constraints they must satisfy. An art proceeds by heuristic, rules of thumb, and "intuition" to search for new things that meet certain goals, and at the same time meet the constraints of reality, the laws of the relevant underlying sciences. No gravity shields; no perpetual motion machines.

For many years after World War II, science was steadily replacing design in engineering college curricula, for we knew how to teach science in an academically respectable, that is, rigorous and formal, way. We did not think we knew how to teach an art. Consequently, the drawing board disappeared from the engineering laboratory—if, indeed, a laboratory remained. Now we have the beginnings—more than the beginnings, a solid core—of a science of design.

One of the great gifts of the modern computer has been to illuminate for us the nature of design, to strip away the mystery from heuristics and intuition. The computer is a machine that is capable of doing design work, but in order to learn how to use it for design, an undertaking still under way, we have to understand what the design process is.

We know a good deal, in a quite systematic way, about the rules of thumb that enable very selective searches through enormous spaces. We know that "intuition" is our old friend "recognition," enabled by training and experience through which we acquire a great collection of familiar patterns that can be recognized when they appear in our problem situations. Once recognized, these patterns lead us to the knowledge stored in our memories. With this understanding of the design process in hand, we have been able to reintroduce design into the curriculum in a way that satisfies our need for rigor; for understanding what we are doing and why.

One of the authors of this book is among the leaders in creating this science of design and showing both how it can be taught to students of engineering and how it can be implemented in computers that share with human designers the tasks of carrying out the design process. The other is leading the charge to integrate the management sciences into both engineering education and the successful conduct of engineering design projects. This book thus represents a marriage of the sciences of design and of management. The science of design continues to move rapidly forward, deepening our understanding and enlarging our opportunities for human-machine collaboration. The study of design has joined the study of the other sciences as one of the exciting intellectual adventures of the present and coming decades.

Herbert A. Simon
Carnegie Mellon University
August 6, 1998

Preface

Why should you read this book on engineering design?

There have been many books written on design, engineering design, project management, team dynamics, project-based learning, and the other topics we cover in this volume. (Check out our representative bibliography if you doubt that!) This book grew out of our own need for a text that combines all of these topics, with a particular focus on methods and techniques for conceptual design. That need arose in our teaching at Harvey Mudd College, where our students do team-based design projects in a freshman design course ("Introduction to Engineering Design," which we affectionately call "E4") and in the Engineering Clinic. Clinic is an unusual capstone "course" taken during junior (for one semester) and senior (for both semesters) years in which students work on externally sponsored design and development projects. In both E4 and Clinic, Mudd students work in multidisciplinary teams, under specified time deadlines, and within specified budget constraints. These conditions replicate to a significant degree the environments within which most practicing engineers will do much—or most—of their professional design work. In looking for books that could serve this audience, we found that there were excellent texts covering detailed design, usually targeted toward senior capstone design courses, or introductions to engineering that focused on detailing the branches of engineering rather than helping students get started in doing engineering design. We could not find a book that introduced the processes and tools of conceptual design in a project or team setting without presuming substantial engineering knowledge.

This book is really directed to three related audiences: students, teachers, and practitioners. While each group has its own special concerns, those interests are intimately tied together by the topics of this book.

We hope the book will enable *students* to learn the nature of design, which is certainly the most central activity of engineering, either directly or as an ultimate aim. (Even the most focused materials engineering scientists, for example, hope that any new materials they develop will be used in the design of a *thing*.) We also hope the book will help students learn the tools and techniques of formal design that will be useful in framing the design problems they will face during their education and during their careers. We are certain the issues and tools of project management will also be used by students when they go on to careers in industry, government, and even in the academy. Since design (and research) are increasingly done in teams, the insights and tips on team dynamics will be valuable as young engineers reflect on their occupational life as well. We have included examples of work done by our students on actual projects for our E4 course, both to show how the tools are used and to highlight some frequently made mistakes. We hope student readers will take heart from these illustrations and use them in learning to be effective engineers (or at least effective engineering students).

The book was written also with *teachers* in mind. This is almost inevitable, given our occupations as professors at an undergraduate college of engineering and science, but we have tried to be explicit in our consideration of how to deliver this material to students and of how professors might want to teach introductory design courses. To this end, this book is structured to allow the teacher to use ongoing examples for illustration and as homework or in-class exercises. The ordering of the material is such that professors can decide for themselves whether to cover the ideas in the text prior to or concurrently with particular stages in design projects. We have tried both approaches in our courses and find that both have their benefits. In an accompanying *Instructor's Manual* we present sample syllabi that detail organizations for teaching the material in the book, as well as numerous additional examples. We also discuss the structure of the material in the book later in the preface to help teachers decide how to use the book.

Finally, we hope the book will be useful to *practitioners*, either as a refresher of things learned or as an introduction to some essential elements of conceptual design that were not formally introduced in engineering curricula of years past. We do not assume that the examples given here substitute for an engineer's experience, but do believe that case studies such as that found in Chapter 10 show the relevance of these tools to practical engineering settings. Some of our friends and colleagues in the profession like to point out that the tools we teach would be unnecessary if only we all had more common sense. Notwithstanding that, the number and scale of failed projects suggests that common sense may not, after all, be so commonly distributed. In any case, this book offers both practicing engineers (and engineering managers) a view of the design tools that even the greenest of engineers will have in their toolbox in the coming years.

OUR EXPERIENCE AT HARVEY MUDD

The E4 design course, introduced at Harvey Mudd College in its current form in spring of 1992, is structured to introduce students to three basic elements of contemporary design projects: use of formal design methods, management of small projects with hard deadlines, and aspects of group dynamics. Students learn this by completing conceptual designs for artifacts or devices such as a corn degrainer for use in developing countries, a giant calculator for use in math programs, and a Braille printer for use with personal computers. In each design project, our student teams are required to use the formal design tools to perform their design effort. Clearly, not all projects find every tool equally useful. This is analogous to the situation faced by an apprentice carpenter. She may be asked to use each tool in her carpenter's toolbox, and she may conclude that some of them are not as easy to use, or as useful in all situations, as others. In E4 we ask only that the students use the design tools reflectively and then document the outcomes of their usage.

Some 35 years ago, Harvey Mudd College also pioneered the Engineering Clinic experience. Engineering Clinic is a program modeled on the medical idea of clinical practice. As mentioned earlier, Clinic students work in teams on industrially sponsored design and development projects in their junior (for one semester) and senior (both semesters) years. Student teams work closely with a liaison designated by the sponsor, and they make formal and informal oral presentations throughout the year, and deliver written proposals and mid-year and final reports. The Clinic experience represents a logical extension of using a project-based approach to engineering design as a central theme in engineering education—and the approaches developed in this book are also widely and usefully applied in those Clinic design experiences.

We, and our colleagues at Mudd, believe that design should be a foundation or *cornerstone* of engineering education, not just the capstone confined to a senior design project. We believe that design should be taught from the very beginning of an engineering curriculum, and throughout its four years. From our many years of experience we know that engineering freshmen are capable enough and interested enough to put together components, match them in a systems-like approach, recognize performance characteristics, and link components accordingly. We have seen ample evidence that first-year students can: ask a client about objectives and translate them into performance specifications; learn to construct morphological charts to represent their design spaces; and compare different performance metrics to make choices that cannot be dictated by—or even expressed in—formulas from calculus or physics.

SOME SPECIFICS ABOUT WHAT'S COVERED

The organization of this book is guided by our beliefs about design, namely, that it is an *open-ended* and *ill-structured* process. As we discuss in several early chapters, designers have to provide an orderly process for organizing the design activity in order to support making decisions and tradeoffs among competing possible solutions. As such, algorithms and mathematical formulations cannot replace the imperative to understand the needs of various stakeholders (clients, users, the public, and so on), even if those mathematical tools may be useful later in the design process. This lack of structure and available formal mathematical tools make the introduction of conceptual design early in the curriculum possible and, we would argue, desirable. It provides a framework in which engineering science and analysis can be used, while not demanding skills that most first- or second-year students have not acquired. To this end, we have included in this book the following specific tools for conceptual design, for acquiring and organizing design knowledge, and for managing the team environment in which design takes place.

The following *formal conceptual design methods* are discussed:

- objectives trees
- pairwise comparison charts
- weighted objectives trees
- function–means trees
- functional analysis
- requirements matrices
- performance specification method
- morphological charts (or, function–means tables)

Since conceptual design thinking both requires and produces a great deal of information, we also discuss the following *means of acquiring and processing information*:

- literature reviews
- brainstorming
- synectics and analogies
- user surveys and questionnaires
- benchmarking
- reverse engineering (dissection)
- definition of metrics
- laboratory experiments
- simulation and computer analysis
- formal design reviews

Since the successful completion of any design project by a team requires that team members estimate a project's scope of work, schedule, and resources early in the life of the project, we describe several *design management tools*, including:

- work breakdown structures
- linear responsibility charts
- schedules
- activity networks
- Gantt charts
- budgets
- control tools

OUR USE OF INTEGRATIVE DESIGN EXAMPLES

One of the features of our text that we like—one that we think is rare, if not unique—is that we use several integrative examples that follow the design process consistently through to completion, thus showing each of the tools and techniques in actual use on the same design project. In addition to numerous "one-time" examples, we describe the following integrative examples:

1. Design of a *beverage container*. The designer, having a fruit juice company as a client, is asked to develop a means of delivering the juice to a market predominantly composed of children and their parents. There are clearly a number of possibilities (e.g., mylar bags, molded plastics), and issues such as environmental effects, safety, and cost of manufacturing are considered.

2. Design of a *chicken coop* to be built and used by a Mayan cooperative in Guatemala. The chicken coop was designed by a team of Harvey Mudd College students in our E4 design class. It was subsequently built by the students, on-site and of indigenous materials, with support from a humanitarian aid group called Xela-Aid.

3. Design of a *transportation network* to enable automobile commuter traffic between Boston and its northern suburbs, through Charlestown, Massachusetts. This conceptual design problem clearly illustrates the many factors that go into large-scale engineering projects in their early stages, when choices are being made between highways, tunnels, and bridges. Among the design concerns are cost, implications for future expansion, and preservation of the character, environment, and even the view of the affected neighborhoods. This project is also an example of how conceptual design thinking can significantly influence some very "real world" events.

We also cover several topics that may be viewed as less central in a first exposure to design, but we feel they are very important. Thus, in Chapter 7 we discuss the endgame and completion of a design project, with a strong emphasis on the ways and means of reporting design results. In Chapter 8 we discuss various "design for X" issues, including manufacturing and assembly, affordability (engineering economics), and reliability and maintainability. In Chapter 9 we discuss various ethical issues in design.

SOME FINAL REFLECTIONS ON THE BOOK AS A PROJECT

When we sat down and actually started writing the book, we confronted many of the same issues that we discuss in the pages that follow. It was clearly important for us to keep in mind our overall objectives for the book, which we have outlined above, as well as the particular objectives we had for each chapter. We had to ask what pedagogic function was served by the various examples, and whether some other example or tool might provide a better means for achieving that pedagogical function. Writing a book is definitely one kind of design project—it requires the same concerns with scope, spending, and schedule as do other engineering or design projects.

Our work as co-authors also gave us firsthand experience with the issues of team dynamics and of conflict resolution. Virtually every chapter was marked by idea-based constructive conflict that led, we think, to a better and more readable book. There were also some moments when our different work styles led to unproductive conflicts. Nevertheless, we emerged from these with a sharper and clearer picture of where we were going and why. This experience of our book as a project has, we believe, made us more effective as teachers of project-based engineering design. We hope it is similarly beneficial to its readers.

Clive L. Dym
Patrick Little
Claremont, California

Acknowledgments

$\mathbf{A}$ book like this does not get written without the faith, support, advice, criticism, and help from many people. We want to thank some of those people, as follows:

Our colleagues and our students at Harvey Mudd College, especially those who have participated in our E4 course since it was introduced in its current form in spring of 1992. *Rich Phillips*, our department chair, for working to make E4 a signature course for the engineering department by teaching it himself several times and by ensuring that most of our colleagues served on an E4 teaching team at least once. *Jim Rosenberg* for exceptional generosity in sharing his experiences and insights from having taught the course several times. *Joe King* for helpfully reviewing our overview of the role of drawings in design. *Michael James Messina* and *Philip Johnson* for doing some of the figures and making heroic contributions to an integrated conceptual design environment. For allowing us to use their design results, the student teams of *Jeannie Connor, Kristina Kubler, Peter Leitzell, J. P. Strozzo,* and *Mark Wang* (Connor et al. 1997) and *Peter Gutierrez, Joey Kimball, Brian Maul, Adam Thurston,* and *Jake Walker* (Gutierrez et al. 1997).

Susan Carlson Skalak (University of Virginia), *Peter R. Frise* (University of Windsor), *Larry G. Richards* (University of Virginia) and *Burt L. Swersey* (Rensselaer Polytechnic Institute) for their suggestions and insights. Particular thanks to Susan, Peter, and Larry for very constructive comments that helped shape our thinking about the final manuscript.

Herbert A. Simon of Carnegie Mellon University for providing one of us (CLD) with much encouragement over the years and for generously penning the foreword to this book.

William J. LeMessurier of William J. LeMessurier Consultants for reviewing Chapter 9 and providing some very helpful insights.

Miriam Dym for designing the cover.

Ted Belytschko, Ed Colgate, Leon Keer, and *Greg Olson* of Northwestern University for inviting CLD to serve as Eshbach Visiting Professor of Civil Engineering at Northwestern during 1997–98, where much of this manuscript was begun; Leon and Ted for handling the administrative stuff, Ed for providing great support while leading the development of a new first-year course ("Engineering Design and Communication"), and Greg for being a stimulating interlocutor on all sorts of design issues.

Joe Hayton of John Wiley & Sons for his unwavering support of this project from its inception.

J. Stanley Johnson and *Mary Wig Johnson* for providing Harvey Mudd with an endowed chair from which to write this book, and, even more significantly, for tasking PL to look for ways to incorporate management into engineering education.

Fred Salvucci of the Massachusetts Institute of Technology for suggesting the City Square project as an example that used the techniques we describe, and for generously describing the effect of the project on his own formation as an engineer. *Andreas Aeppli*, of A & L Associates, for suggesting a number of contacts for exploring the City Square example, and for making office space and resources available when PL was in Boston doing that research. *George Sanborn* of the Massachusetts Transportation Library for finding numerous sources about City Square and the planning processes, and for making a number of those sources available to us.

Joan Dym and *Judy Little*, our wives, for tolerating us during the absences that such a project entails, and for listening to each of us as we worked through our conflicts and differences to find a common voice.

Contents

Contents

5.   Specifications   **108**

How can I express what the client wants in terms that help me as an engineer?

6.   Finding Answers to the Problem   **135**

Which of these ten good ideas is the best answer?

How do we let our client know about our solutions?

What are some of the consequences of our technical choices?

9.   Ethics in Design   217

Isn't design really just a technical matter?

10.   The City Square Transportation Hub: A Design Case Study   233

Can you show me an example from engineering practice?

Chapter 1

Engineering Design

What do you mean when you say you want me to design something? Are engineers who design doing anything different than anyone else who designs things?

People have been designing things for as long as we can "remember" or archaeologically uncover. Our earliest ancestors designed flint knives and other basic tools to help meet their most basic needs. They also designed wall paintings to tell stories and make their primitive caves visually more comfortable. Given the long history of people designing *artifacts*, it is useful to ask whether the engineer designing structural members to support a building's occupants is, somehow, doing the same thing as the interior designer who selects carpets and wall hangings to decorate that building. We will use this chapter to set some contexts for engineering design and to start developing both a vocabulary and a shared understanding of what we mean by engineering design.

1.1 WHERE AND WHEN DO ENGINEERS DESIGN?

What does it mean for an *engineer* to design something? When do engineers design things? Where? Why? For whom?

There are a lot of questions we could ask about engineers doing design, and there are probably more answers than there are questions. An engineer could be working for a large company that processes and distributes various food products, in which case she might be asked to design a new container for a new juice product. He could be working for a design-and-construction company, for which he might be designing some part of a bridge for a new interstate highway that is part of a larger transportation project. Or an engineer might be working for an automobile company that wants to develop a new concept for the instrumentation cluster in its cars, perhaps one that makes it easier for drivers to check various parameters without having to take their eyes off the road. Or the engineer might be working for a school system that wants specialized facilities to better serve students with various kinds of orthopedic disabilities.

Clearly, this is a list that could easily be made very, very long, which makes it worth asking whether there are any common elements either in the engineers' situations or in the ways that they approach their tasks. It turns out that there are common features in both situations and tasks, and it is the existence of such commonalities that makes it possible to describe both the design process and the context in which it occurs.

To begin with, we can identify three "roles" being played as the design of a product unfolds. Obviously there is the *designer*, and it seems equally clear that there will be a *client*, the person or group or company that wants a design conceived. For the working engineer, the client could be internal (e.g., the person who decides the food company should start selling a new juice product) or external (e.g., the government agency that contracts for the new highway system). And while there may be some differences in how the designer relates to internal and external clients, in both cases it is the client who presents a project statement from which all else begins to flow. Design project statements are often verbal, and sometimes they are quite short. These two qualities suggest that the designer's first task is to clarify what the client really wants and translate it into a form that is useful to her as an engineering designer. We'll say much more about this in Chapter 2 and beyond, but we want to recognize for now that a design is motivated by a client who wants some sort of product.

There's another player or stakeholder in the design effort and that's the *user*, the person or set of people who will actually use the device or artifact being designed. In the contexts mentioned above, the users would be consumers who buy the new juice, drivers on the interstate highway system, drivers of the new line of autos, and orthopedically disabled students. The users hold a stake in the design process because a product "won't sell" if its design doesn't meet their needs. Thus, the designer, the client, and the user form an interesting triangle, as we show in Figure 1.1. It is clear that the designer has to understand what the client wants, but the client also has to understand what his users need and help communicate that to the designer. In Chapter 2 we will describe design processes that model how the designer can interact and communicate with both the client and potential users to help inform her own design thinking, and we will identify some tools (that will be discussed in detail in Chapters 3–7) that she can use to organize and refine her thinking.

Engineering designers work in many different kinds of environments, some of which clearly emerged in the contexts described above. It is hard to be very specific about a designer's working environment because it can include companies large and small, including start-up ventures, government and other not-for-profit agencies, and engineering services firms, one breed of which is the industrial design consultancy. Apart from the salaries paid and perks of working in these

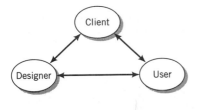

Figure 1.1 The designer–client–user triangle. There are three parties involved in a design effort: the client, who has objectives that the designer must clarify; the user of the designed device, who has his own requirements; and the designer, who must develop specifications such that something can be built to satisfy everybody!

various places to do design, the designer will most likely see differences arising due to the size of the project, the number of colleagues on the design team, and the designer's access to relevant information about user needs. On large projects, many of the designers will be working on pieces of a project that are so detailed and so confined that much of what we describe in this book may not seem immediately useful. Thus, for example, the designers of a bridge abutment, an airplane's fuel tank, or components on a computer motherboard will likely not be as concerned with the larger picture of what external clients and users want. Indeed, as we will explain in Chapter 2, these kinds of design problems are the part of the process called *detailed design*, in which the choices and procedures are so well understood that more general design issues have already been taken into account. However, the response to a client's project statement, even for such large projects, is initiated with *conceptual design*. Some thinking about the size and mission of the airplane will have been done to determine the constraints surrounding the design of fuel tanks, while the performance parameters that the computer motherboard must display will be determined by some assessment of the market for and the price of the computer in question.

Large, complex projects often lead to very different interpretations of client project statements and of user needs. One has only to think of the many different kinds of skyscrapers that decorate our major cities to see how architects and structural engineers envisage different ways of housing people in offices and apartments. More visible differences emerge in airplane design (Figure 1.2) and wheelchair design (Figure 1.3). Each of these two kinds of devices can follow from a simple, common design statement: The airplanes are "devices that safely transport people and goods through the air," while all the wheelchairs are "personal mobile devices that transport people who are unable to use their legs." However, the different products that have emerged represent different conceptualizations of what the clients and users wanted from these artifacts. Designers thus have much to do when clarifying what a client wants and translating those wants into an engineered product.

Our players' triangle also prompts us to consider that the interests of all three participants could diverge, and to recognize that the consequences of such divergence could mean more than financial problems resulting from a failure to meet users' needs. This happens because the interaction of several interests also creates an interaction of multiple obligations, and these obligations may well conflict. For example, the designer of the juice container will likely consider that users' ability to handle containers in different configurations could vary with age, in which case the assessments of the safety of new juice containers will vary

Figure 1.2 A collection of "devices that safely transport people and goods through the air," i.e., airplanes. No surprises here, right? We've all seen a lot of airplanes (or, at least, in pictures or movies). But even these planes, albeit of different eras and origins, show that they were clearly designed to achieve very different missions.

Figure 1.2 Continued.

Figure 1.3 A collection of "personal mobile devices that transport people who are unable to use their legs," i.e., wheelchairs. Here too, as with the airplane, we see some sharp differences in the configurations and components of these wheelchairs. Why are the wheels so different? Why are the wheelchairs so different?

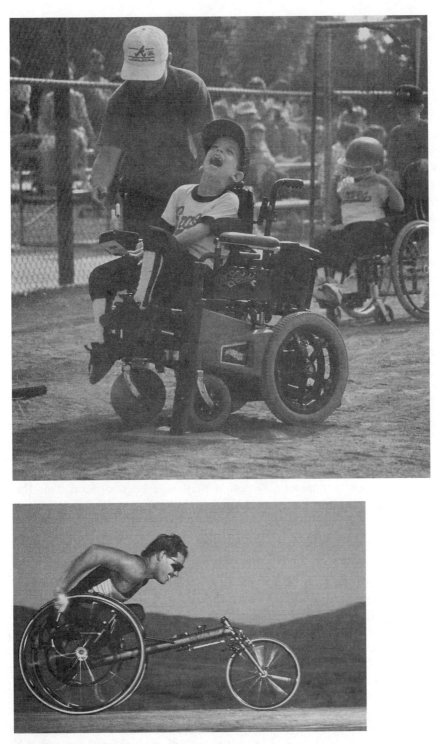

Figure 1.3 Continued.

with market segments that correlate with customer age. For example, a metal can that is easily "squashed" could become a hazard if sharp edges are produced during the squashing. There might be tradeoffs among design variables, including the material of which a container is to be made and the container's thickness. The choices made in the final design could easily reflect different assessments of the possible safety hazards, and this in turn lays the foundation for potential ethics problems. Ethics problems, which will be discussed further in Chapter 9, occur because the designer has, in fact, obligations not just to the client and the user, but also to her profession and, as detailed in the codes of ethics of engineering societies, to the public at large. Thus, ethics issues are never terribly far from the design process.

Another aspect of engineering design practice that is increasingly common in projects and firms of all sizes is the use of *teams* to do design. Many engineering problems are inherently multidisciplinary, for example, the design of medical instrumentation, so there is a need to understand the requirements of clients, users, and technologies in very different environments. This, in turn, requires the assembling of teams that can address such different sets of environmental needs. The widespread use of teams clearly affects the management of design projects, another recurring theme of this book.

Engineering design is clearly a multifaceted subject, and in no way do we want to pretend that we can describe and explain this wonderfully complex activity in just one short book. However, we do hope that we can successfully model ways of thinking about some of the conceptual issues and the resulting choices that are made very early in the design of many different kinds of engineered artifacts. With that in mind, we move now to defining what we mean by engineering design.

1.2 DEFINING ENGINEERING DESIGN

Of all the definitions of engineering design that have been offered over the years—and there are quite a few—we think that the following formal definition of engineering design is the most useful one for our purposes:

> *Engineering design* is the systematic, intelligent generation and evaluation of specifications for artifacts whose form and function achieve stated objectives and satisfy specified constraints.

What precisely does this mean?

First of all, *artifacts* are human-made objects, the "things" or devices that we are designing or creating. These things are most often "physical," like airplanes, wheelchairs, ladders, and carburetors. But artifacts can also be "paper" products, such as drawings, plans, computer software, articles, and books. The *form* of the artifact is its shape, its geometry. By *function* we mean those things the artifact is supposed to do. The *specifications* of artifacts are precise descriptions of the properties of the object being designed. Typically they are numerical values of *performance parameters* (i.e., constants or variables that serve as indicators of the artifact's behavior) or of *attributes* (i.e., properties or characteristics of the artifact). For example, the various lengths associated with an extension ladder could include both geometric attributes and performance parameters. Thus, a "folded," unextended length of five feet would be an attribute of the ladder, while the requirement that the ladder permits access to heights of ten feet would be a performance specification. As a matter of terminology, we will refer to *design specifications* as that set of values that articulates what a design is intended to do. Design specifications, then, also provide a basis for *evaluating* proposed designs, as they become the "targets" of the design process against which we can measure our success in achieving them.

The definition states the design specifications are found as the result of *systematic, intelligent generation*. This is not to say that design is not a creative process, for it is. However, it is also true that there are techniques and tools we can use to support our creativity, to help us think more clearly, and make better decisions along the way. These tools and techniques, which form much of the subject of this book, are not formulas or algorithms. Rather, they are ways of asking questions, and presenting and viewing the answers to those questions as the design process unfolds. Along the way we will also present some tools and techniques for managing a design project. Thus, in presenting the design tools we will be demonstrating ways of thinking about a design as it unfolds in our heads. In presenting the design management tools we will be demonstrating ways of organizing the resources needed to complete a design project without exceeding budgets of either time or money.

An *objective*, to quote Webster's dictionary, is "something toward which effort is directed: an aim, goal, or end of action." Similarly, a *constraint* is a "condition, agency, or force" that imposes a "stricture, restriction, or limitation." Thus, if we were designing a corn degrainer that could be cheaply built of indigenous materials by Nicaraguan farmers, we might have as an objective that it should be as cheap as possible, while we might impose the constraint that under no circumstances could it cost more than $20.00 (US). The statement that the degrainer be made of indigenous or local materials could be viewed either as an objective or a constraint, so that distinguishing one from another might depend on whether it is a desired feature, in which case we're talking objectives, or an absolutely required feature, in which case we're talking constraints.

Having explained what our technical terms mean, we can now rewrite our definition of engineering design in more colloquial terms. Thus,

> *Engineering design* is the organized, thoughtful development and testing of characteristics of new objects that have a particular configuration or perform some desired function(s) that meets our aims without violating any specified limitations.

There are some implicit assumptions captured in this definition (as well as in its formal counterpart), and it will be very helpful for us to make them explicit.

First of all, design is a *thoughtful* process that can be *understood*. Thus, without meaning to spoil the magic (and the importance) of creativity in design, we do note that people *think* while designing. We want to provide tools to support that thinking, that is, to support design decision making and design project management. Incidentally, one reason we know that people think while they're designing is that we have been able to simulate design processes on computers—which is something we couldn't do if we couldn't articulate and describe what goes on in our heads when we design things.

Strongly related to our inclination to think about design is the idea that there are *formal methods* that we can use to generate design alternatives. This might seem pretty obvious because there's not much point in looking for new ways of looking at design problems or talking about them—unless we can exploit them to do design more effectively. This may also distinguish an "engineering" approach from a more "scientific" approach, because we are interested here in the usefulness of new ways of talking about or representing design issues.

We can talk about and describe both *form* and *function* as two related but independent entities. This is important because we often think of the design process as beginning when we sit down to draw or sketch something, which suggests that form is a typical starting point. However, while we focus on form, we need to keep in mind that function is an altogether different aspect of a design that may not have an obvious relationship to its shape or form. In particular, while we can often infer the purpose of an object or artifact from its form or structure, we can't do the reverse, that is, we can't *automatically* deduce what form an artifact must have *from the function alone*. For example, we can look at a pair of connected boards and deduce that the devices that connect them (e.g., nails, nuts and bolts, rivets, screws, etc.) are fastening devices whose purpose or function is to connect the individual members of each pair. However, if we were to start with a statement of purpose that we wish to connect two boards, there is no obvious link or inference that we can use to create a form or shape for a fastening device.

We also need to remember that the endpoint of a successful design is the production of a set of plans for making the designed artifact. This set of plans, called *fabrication specifications*, must be clear, unambiguous, complete, and transparent. That is, the fabrication specifications must, on their own, make it possible for someone totally unconnected to the designer or the design process to make or fabricate what the designer intended in such a way that it performs just as the designer intended. This is a facet of modern engineering practice that represents a sharp departure from "the good old days" when designers were often craftsmen who made what they designed. These designer-fabricators were allowed some latitude or shorthand in their designs because, as fabricators, they knew exactly what the designer intended. In more recent times it has been rare that engineers make what they design. Rather, designs were typically "thrown over the wall" to a manufacturing department or to a contractor—and all the fabricator or contractor knew was "what's in the specs." Current practice has evolved with regard to this issue, and we'll discuss that in the next section.

There are criteria that we can use as benchmarks to evaluate our progress toward a design, as well as the performance of the final design itself. These evaluation benchmarks are *design specifications* or *requirements* that result from the designer's *translation* of the client's objectives into goals for the artifact being designed. Design specifications are stated in a number of ways, depending on the nature of the requirements the designer wants to articulate.

Prescriptive specifications are used to specify values for attributes of the designed object. For example, "A step on a ladder is safe if it is made from Grade A fir, has a length that does not exceed 20 in., and is attached in a full-width groove slot at each end."

Procedural specifications are used to identify specific procedures for calculating attributes or behavior. For example, "A step on the ladder is safe if its maximum bending stress is computed from $\sigma_{max} = \frac{Mc}{I}$ and is such that σ_{max} does not exceed σ_{allow}."

Performance specifications are used to characterize the desired behavior. For example, "A step on a ladder is safe if it will support an 800 lb gorilla."

We might also remember here that it is often the case that the manufacture or use of a device points up deficiencies that were not anticipated in the original design. That is, successful designs often produce unanticipated secondary or tertiary effects which can become *ex post facto* evaluation criteria. For example, because of its contribution to air pollution and other *unintended consequences*, some people have regarded the automobile as a failure. Perhaps a different way of looking at the automobile—which does, after all, provide the means of personal transportation that it was intended to—is to note that changing societal expectations have dictated serious redesign of many of its attributes.

Finally, our definition of engineering design and the related assumptions we have identified clearly rely heavily on the notion that communication is central to the design process. This means that some set of languages or representations is inherently and unavoidably involved in every part of the design process. From the original communication of a design problem, through its translation into design specs and an eventual solution, to its evaluation and fabrication, the artifact being designed must be described; it must be "talked about." Thus, *communication is the key issue*. It is not that problem solving and evaluation are less important; they are extremely important, but they too must be expressed and implemented at levels and in styles—in a spoken or written language, in numbers, in equations, in rules, in charts, in pictures—that are appropriate to the immediate task at hand. Thus, successful work in design is inextricably bound up with the ability to communicate.

1.3 MORE ON DESIGN AND ENGINEERING DESIGN

We've noted that people have been studying design—as well as doing it and talking and writing about it—for quite a while now. Yet it is still not all that easy to make sense of design to someone who wants to learn *how to do it*. Like riding a bicycle or throwing a ball, like drawing and painting and dancing, it often seems easier to say to a student, "Watch what I'm doing and then try to do the same thing yourself." Clearly, there's a *studio* aspect to trying to teach any of these activities, an element of *learning by doing*.

One of the reasons that it isn't easy to teach someone how to do design—or to ride a bike or throw a ball or draw or dance—is that we're often better at *demonstrating* a skill than we are at *articulating* what we do know about applying our various skills. Some of the skills just mentioned clearly involve some physical capabilities, but the difference that is of most interest to us is not simply that some people are more gifted physically than others. What is really interesting is that someone who throws a softball cannot tell you just how much pressure she exerts when holding the ball, nor how fast her hand ought to be going, or in what direction, when she releases it. Yet, somehow, and almost by magic, the softball goes where it's supposed to go and winds up in the hands of a catcher. The real point is that the thrower's nervous system contains some knowledge that allows her to assess distances and choose muscle contractions to produce a desired trajectory. And while we can model that trajectory, given initial position and velocity, we don't yet have the ability to model the knowledge in the nervous system that generates that initial data.

The language of mathematics has been extraordinarily useful over quite some time for building models of phenomena in the physical sciences, such as the trajectory model just mentioned, and of devices and their behavior. However, only recently have we begun to develop a similar facility with a language or set of languages that we can use to describe how we actually think about artifacts and how those designed objects came to be. It's not that people weren't thinking while designing, but that we couldn't model their thought processes. For example, with regard to the design of early elementary artifacts, it is almost certainly true that the "designing" was inextricably linked with the "making" of these primitive implements. There is no record of a separate, discernible modeling process. However, we can never know for sure, because who is to say that small flint knives, for example, were not consciously used as models for larger, more elaborate cutting instruments? Certainly people must have *thought* about what they were making as they recognized shortcomings or failures of devices already in use and evolved more sophisticated versions of particular artifacts. Even the simple enlargement of a small flint knife to a larger version could have been driven by the inadequacy of the smaller knife for cutting into the hides and innards of larger animals. But we really have no idea of how these early designers thought about their work, what kinds of languages or images they used to process their thoughts about design, or what mental models they may have used to assess function or judge form. If we can be sure of anything, it is that much of what they did was done by trial and error. (The process in which trial solutions are generated by unspecified means and then tested against given evaluation criteria is now called *generate and test*.)

What we do know about design problems in general—and engineering design problems in particular—is that they are said to be *open-ended* and *illstructured*. We say that design problems are *open-ended* because they usually have several acceptable solutions. The quality of uniqueness, so important in many mathematics and analysis problems, simply does not apply. We also say that design problems are *illstructured* because their solutions cannot normally be found by applying mathematical formulas or algorithms in a routine or structured way.

We can see evidence for these two characterizations in a familiar household object, such as the ladder. We show several ladders in Figure 1.4, including a stepladder, an extension ladder, and a rope ladder. Unless and until we determine a specific set of uses for a ladder, we can't identify a particular ladder design to target. Even if we've determined that a particular form is appropriate for the household handyman, say, a stepladder, there are other questions that come to mind. Of what materials might it be made? Wood, aluminum, plastic, or a fancy high-tech composite? How much should the ladder cost? And, can we say which ladder design would be the *best*? Is it possible to identify the *best* ladder design, the *optimal design*? Actually, the answer is, "No," we can't stipulate a single or universally optimal ladder design, although we might be able to optimize a ladder design under some very tightly constrained circumstances.

Figure 1.4 A collection of "devices that enable people to reach heights they would be otherwise unable to reach," that is, ladders. Note the variety of ladders, from which we can infer that the design objectives involved a lot more than the simple idea of getting people up to some height. Why are these ladders so different?

Figure 1.4 Continued.

How do we talk about all of these issues, for example, purpose, intended use, materials, cost, and possibly other concerns? In other words, how do we articulate the choices about both the form and function of the ladder? Clearly, there are different ways of representing these differing characteristics by using various "languages." Even the simple ladder design problem becomes a complex study that shows the two characteristics—poorly defined endpoints and ill-defined structure—that make design such a tantalizing and interesting subject. How much more complicated and interesting are projects to design a new automobile, a skyscraper, or a way to land someone on the moon.

We find examples of works that must have been designed, such as the Great Pyramids of Egypt, the Great Wall of China, and the cities and temples of Mayan civilization, well back in recorded history. Unfortunately, the designers of these wonderfully complex structures did not leave a "paper trail" that recorded their thoughts on their designs. However, there are some discussions of design that go pretty far back, one of the most famous being the collection of works by the Venetian architect Andrea Palladio (1508–1580). Palladio's works were apparently first translated into English in the eighteenth century, in which language they are still available today. Since then, discussions of design have been developed in fields as diverse as architecture, organizational decision making, and various styles of professional consultation, including the practice of engineering.

There are a lot of words that appear often in discussions of design, including, for example, form, function, specification, and optimum. One author observed, however, that in one list of eleven definitions of design, only about one-tenth of the really "important words" appeared more than once in that entire list! Thus, it should not surprise us that there are many, many definitions of design, or that these definitions sound like they're trying to say similar things in different words. This is due in part to the fact that it is as difficult to define design generally, as it is to define the particular endeavor of engineering design. For example, we might observe that design is a goal-directed activity, performed by humans, and subject to constraints. The product of this design activity is a *plan* to realize those goals.

Herbert A. Simon, Nobel laureate in economics and founding father to several fields, including design theory, has offered a definition that is closely related to our engineering concerns. As an activity, design is intended to produce a "description of an artifice in terms of its organization and functioning—its interface between inner and outer environments." Designers are thus expected to describe the shape and configuration of a device (its "organization"), how that device does what it was intended to do (its "function"), and how the device (its "inner environment") works ("interfaces") within its operating ("outer") environment. Simon's definition is interesting for engineers because it places designed objects in a *systems* context, wherein we recognize that an artifact must operate as part of a system that includes the world around it.

We noted earlier that primitive designers almost certainly designed artifacts through the very process of making them. It has been said that this approach to design is a distinguishing feature of a *craft*. With this in mind, it is also useful for us to distinguish engineering design from those domains—or crafts—wherein the designer actually produces the artifact directly, such as graphics or type design. We have already observed that engineering designers do not, typically, produce artifacts; rather, they produce the fabrication specifications for making artifacts. That is, the designer in an engineering context produces a detailed description of the designed device so that it can be assembled or manufactured, thus separating the "designing" from the "making." And we have noted that the specification must be both complete and quite specific; there should be no ambiguity and nothing can be left out.

Traditionally, fabrication specifications have been presented through some combination of drawings (e.g., blueprints, circuit diagrams, flow charts, etc.) and text (e.g., parts lists, materials specifications, assembly instructions, etc.). We can achieve completeness and specificity with traditional specifications, but we cannot capture in them a designer's intent—and this can lead to catastrophe. In 1981, a suspended walkway in the Hyatt Regency Hotel in Kansas City collapsed because a contractor fabricated the connections for the walkways in a manner different than intended by the original designer. In that design, walkways at the second and fourth floors were hung from the same set of threaded rods that would carry their weights and loads to a roof truss (see Figure 1.5). The fabricator was unable to procure threaded rods sufficiently long (i.e., 24 ft) to suspend the second-floor walkway from the roof truss, so instead, he hung it from the fourth-floor walkway with shorter rods. (It also would have been hard to screw on the bolts over such lengths and attach walkway support beams.) The fabricator's redesign was akin to requiring that the lower of two people hanging independently from the same rope change his position so that he was grasping the feet of the person above, and that person was then carrying both their weights

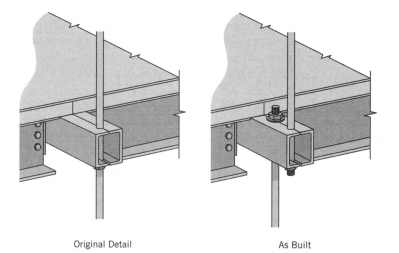

Original Detail   As Built

Figure 1.5 The walkway suspension connection—as originally designed and as built—in the Regency Hyatt House in Kansas City. We see that the change made during construction resulted in the second-floor walkway being hung from the fourth-floor walkway, rather than being connected directly to the roof truss.

with respect to the rope. In the hotel, the supports of the fourth-floor walkway were not designed to carry the second-floor walkway in addition to its own dead and live loads, so a collapse occurred, 114 people died, and millions of dollars of damage was sustained. If the fabricator had understood the designer's intention to hang the second-floor walkway directly from the roof truss, this accident might never have happened. Thus, had there been a better way for the designer to explicitly communicate his intentions to the fabricator, a great tragedy might have been avoided.

We also learn here another consequence of separating the "making" from the "designing." If the designer had worked with a fabricator or a supplier of threaded rods while he was still designing, he would have learned that no one made threaded rod in the lengths needed to hang the second-floor walkway directly from the roof truss. Then the designer would have sought another solution in an early design stage. In the manufacturing sector of our economy, it was typically the case that there was a "brick wall" between the design engineers and the manufacturing engineers. Only recently has this wall been penetrated—even broken. Manufacturing and assembly considerations are increasingly addressed during the design process, rather than afterward. One element in this new practice is *design for manufacturing*, in which the ability to make or fabricate an artifact is specifically incorporated into the design specifications, perhaps as a set of manufacturing constraints. The idea here is that the designer must be aware of parts that are difficult to make or of limitations on manufacturing processes as a design unfolds. A second new idea is *concurrent engineering*, wherein designers, manufacturing specialists, and designers concerned with the product's life cycle (e.g., purchase, support, use, and maintenance) work together, along with others who hold a stake in the design, so that they are collectively and concurrently designing an artifact together. Concurrent engineering demands teamwork of a high order. Its practice has resulted in the replacement of the "over the wall" practices that we mentioned in Section 1.2. Research in this area focuses on ways to enable teams to work together on complex design tasks when team members are dispersed not only by engineering discipline, but also geographically and by time zones.

The Hyatt Regency tale and the lessons we drew suggest that the statement or representation of fabrication specifications is really important. We have learned from this disaster that it is essential that fabrication specifications be complete and specific (or lacking ambiguity), and such that the designer's intent must be clearly communicated. We could also use these specifications as a basis for evaluating how well a design meets its original design goals. Thus, we want to be able to translate the original objectives (and constraints) of the client into some specifications, and to recognize that these same specifications provide the starting point for manufacturing and for design evaluation. A definition of design that explicitly talks about specifications in design is:

> The purpose of design is to derive from a set of specifications a description of an artifact sufficient for its realization. Feasible designs not only satisfy the specifications, but take into account other constraints in the design problem arising from the medium in which the design is to be

executed (e.g., the strength and properties of materials), the physical environment in which the design is to be operated (e.g., kinematic and static laws of equilibrium) and from such factors as the cost and the capabilities of the manufacturing technology available.

This last definition of engineering design begins to focus more on the object being designed, rather than on the cognitive process of thinking about the design. It also brings us much closer to implementing our preferred definition. But while worrying about specifications, as well as all the other issues we have raised, we should also keep in mind that design is a human activity, a social process. This means that communication among and between stakeholders remains a pre-eminent concern. We hope that we have communicated here not only a practical definition of engineering design, but one broad enough to encompass a great variety of concerns.

1.4 MANAGING ENGINEERING DESIGN

It is probably clear by now that a successful design isn't something that just happens. Rather, it is the result of careful thought about what clients and users want and demand, and the specification of ways to realize those requirements. In the coming chapters we will be looking at various tools and techniques that assist the designer in this process. A particularly important element of successful design is *managing* the design project. Just as thinking about design in a rigorous way doesn't imply any loss in creativity, using formal (and informal) tools to manage the process doesn't mean that we give up either technical competency or inventiveness. On the contrary, we can see many examples of organizations that foster imaginative engineering design as an integral part of their management style. At 3M, for example, each of the more than 90 product divisions is expected to generate 25 percent of its annual revenues from products that didn't even exist five years earlier. In this text, we will introduce some of the tools and techniques of management that are most applicable to design projects.

Just as we began by defining terms and developing a common vocabulary for design, we will do the same for management, project management, and the management of design projects. We will go from the general to the specific in these definitions, and in future chapters will look less at definition and more at "hands on" matters. For now, we define management as follows:

> *Management* is the process of achieving organizational goals by engaging in the four major functions of planning, organizing, leading, and controlling.

This definition reflects an emphasis on process and on achieving organizational goals. In this sense, management has something in common with design insofar as both are goal directed and can be considered in terms of steps or processes. We will carry this analogy a little further in Chapter 2 when we consider the phases or stages of design.

The four functions of management can also be defined and discussed in ways that help us see how management might relate to design.

> We define *planning* as "the process of setting goals and deciding how best to achieve them." In this regard, planning involves considering the mission, or ultimate aim, of the organization, and then translating it into appropriate strategic and tactical goals and objectives for the organization.
>
> *Organizing* is "the process of allocating and arranging human and nonhuman resources so that plans can be carried out successfully." Put another way, the organizing function of management is concerned with "creating a framework for developing and assigning tasks, obtaining and allocating resources, and coordinating work activities to achieve goals."
>
> *Leading* is the process of influencing others to engage in behaviors necessary to reach organizational goals. This notion of leadership as a consequence of influence is particularly important in design settings, where a number of different types of influence can come into play. For example, one member of the design team may have influence related to position (e.g., the team leader), while another may have influence based on the team's recognition of her expertise in a particular domain. Leading is the ongoing activity of exerting influence and using power to motivate others to work toward accomplishing goals.
>
> Finally, *controlling* is the process of monitoring and regulating the organization's progress toward achieving goals. Many people find it difficult to understand the difference between leading and controlling, perhaps because we use the term "controlling" as a less-than-flattering synonym for using power in some settings. Since engineers use the term control to refer to monitoring and regulating in many contexts, the difference between leading and controlling seems simpler. In any event, when we talk about control in this book, it means that our primary concern is ensuring that the actual performance conforms to expected standards and goals.

Project management is the application of these four functions (i.e., planning, organizing, leading, and controlling) to accomplish the goals and objectives of a project. A project is "a one-time activity with a well-defined set of desired end results." Examples of engineering projects abound, ranging from construction of new highways (civil engineering), to the development of new computer memories (electrical engineering), to the establishment of processes to be followed on a factory floor (industrial engineering). In each case, however, the common thread is that the project is well defined in terms of its goals, has finite resources, and is to be accomplished in a fixed time frame (sometimes simply as soon as possible). To help project managers in carrying out the four functions, a number of standard tools and techniques have been developed. These include tools for understanding and listing the work to be done; scheduling the tasks to be done logically and efficiently; assigning tasks to individuals; and monitoring progress. We will explore some of these tools and techniques as they are most applicable to design projects later in the book.

Careful readers will note that the precision in goals and objectives that is spoken of regarding projects is somewhat at odds with some of the previous discussion of the open-ended nature of design activities. This is certainly the case when we try to predict the final form or outcome of a design project. Unlike a construction project, where the desired and expected results are clear and generally well articulated, a design project, and especially a conceptual design project, may have a number of possible successful outcomes, or none! This makes the task and tools of project management only partially useful in design settings. As a result, we will only present project management tools that we have found to be useful in managing design projects conducted by small teams.

In addition to a more restricted set of tools for managing the design project, we will be introducing formal tools for guiding the design process itself. These tools are also a form of project management, as they help the team to understand and agree upon goals, organize their activities, organize resources to realize the goals, and monitor whether the alternatives they generate and ultimately select are consistent with their objectives.

1.5 THE ILLUSTRATIVE EXAMPLE APPROACH

Design is best learned both by thinking and by doing. In the words of the famous economist, John Maynard Keynes, "Nothing is required and nothing will avail, except a little, a very little, clear thinking." However, it is also fair to say that *design is best experienced by doing*. To that end, we strongly encourage fledgling designers and engineers to participate actively in teams doing design projects. In addition, some of the formal techniques can be learned by doing exercises and by observing how others have applied the techniques. To this end, throughout the book we will elaborate three examples that illustrate the sorts of problems that engineers face when doing conceptual design, that is, when they have to come up with concepts or ideas to solve a given design problem. These illustrative examples are:

1. Design of a container to deliver a children's beverage. This is a stylized industrial design project that highlights some of the early questions that must be addressed before the engineering designer can apply more conventional engineering science knowledge to the problem.

2. Design of a chicken coop to be built and used by a Mayan cooperative in Guatemala. The chicken coop was designed by a team of Harvey Mudd College students in a freshman design class. It was subsequently built by the students, on-site and of indigenous materials, with support from a humanitarian aid group called Xela-Aid.

3. Design of the City Square transportation network and hub to enable automobile and rail commuter traffic between Boston and its northern and western suburbs, through Charlestown, Massachusetts. This conceptual design problem clearly illustrates the many factors that go into large-scale engineering projects in their early stages, when choices are being made between tunnels or bridges. Among the design concerns are cost, implications for future expansion, and preservation of the character, environment, and even the view of the affected neighborhoods. This project is also an example of how conceptual design thinking can significantly influence very "real world" events.

1.6   NOTES

Section 1.2: Our preferred definition of engineering design appears in (Dym and Levitt 1991) and (Dym 1994).

Section 1.3: Jones (1981) made the observation about the number of "important words" in definitions of design. Simon's definition of design is based on a set of lectures that were published as *The Sciences of the Artificial* (1981). The final definition of design presented is from (Mostow 1987), while it was Leifer who propounded the notion that engineering is a social activity (1991).

Section 1.4: Definitions of management and the four major functions of planning, organizing, leading, and controlling are found in (Bartol and Martin 1994) and (Bovee et al., 1993). The project is defined in (Meredith and Mantel 1995).

Section 1.5: The HMC Xela Aid project results are detailed in (Gutierrez et al., 1997) and (Connor 1997). The City Square project is described in (Barton-Aschman 1962), (Massachusetts Department of Public Works 1974), and (Berger 1981).

1.7   EXERCISES

1.1 Four definitions of engineering design have been offered in this chapter. Identify the key concept in each definition and explain how these four concepts differ.

1.2 List at least three questions you would ask if you were, respectively, a user (purchaser), a client (manufacturer), or a designer who was about to undertake the design of a portable electric guitar.

1.3 List at least three questions you would ask if you were, respectively, a user (purchaser), a client (manufacturer), or a designer who was about to undertake the design of a greenhouse for a tropical climate.

1.4 All aspects of management may be said to be goal directed. Explain how this description is exemplified for each of the four functions of management identified in Section 1.4.

Chapter 2

The Design Process

Can you give me some sort of overview—a road map—of where this book is going?

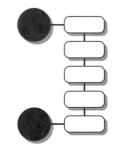

Now that we have defined engineering design and a few technical terms, we go on to explore design as an activity, that is, to explore the *process* of design. Parts of the discussion may seem abstract because we are trying to describe a very complex process by breaking it down into smaller, more detailed pieces. As a way of anchoring the tools we will describe in Chapters 4–7, we will also identify those places in the design process where particular tools or methods can be useful. Please keep in mind, however, that we are not presenting a recipe to be followed in order to complete a design. Rather, in this chapter we are trying to describe *what is going on in our heads when we are doing design.*

2.1 HOW A DESIGN PROCESS UNFOLDS

Suppose we are asked to design a safe ladder. We know that many safe ladders have already been designed, produced, and sold. Indeed, we have seen several different kinds of safe ladders in Figure 1.4. So, what does it mean to start another "safe ladder" project? We could start by asking, what does it mean for a ladder to be "safe"? That it should not tip on level ground? That it should not tip on a mild slope? What is a mild slope? Who, ultimately, will define what "safe" means? Is safety a legal issue or a technical issue?

There are many other questions that arise as we think about the design of a ladder. A partial list could include, in the order that they're likely to come up:

- How is the ladder to be used?
- How high should someone on the ladder be able to reach?
- How much weight should a safe ladder support?
- Should the ladder be portable?
- How does OSHA[1] define safety?
- How much should it cost?
- Is there a market for this ladder?
- Will the ladder be made of wood, aluminum, or fiberglass?
- How many steps are there on the ladder?
- How are the steps to be attached to the frame?
- What is the "allowable load" on a step?
- What is the maximum stress in a step supporting the "design load"?
- How does the bending deflection of a loaded step vary with the material of which the step is made?
- Can the ladder carry the "design load"?
- Can this ladder be assembled?
- Is the design economically feasible?
- Is the ladder safe as it is actually designed?
- Is there a more economic design?
- Is there a more efficient design (e.g., less material)?

There are clearly a lot of questions in this seemingly simple design problem, including others not listed that we might or should have asked. We don't know where the questions came from, whether they were asked by members of the designer–client–user triangle, or whether the questioners were acting singly or as teams. Asking questions to elaborate the meaning of a client's statement is part of the design process, and it is often done by design teams. That's why it is also important to think about the questions themselves we need to ask, while at the same time asking ourselves who else may have useful questions (and answers).

For now we note that most of these questions can't be answered by applying our usual mathematical models of physics. We could probably use Newton's equilibrium law and elementary statics to analyze the stability of the ladder under given loads on a specified surface, and we could write beam equations to calculate deflections and stresses in the steps as they bend under the given loads. But are there equations we can use to define the meaning of "safe," or of marketability, or to choose the ladder's color? No, there aren't. And, since we don't have equations for safety, color, marketability, or for most of the other issues in the list of ladder questions, we need to find other ways to think about this design problem—beyond what we can put into some set of formulas or a computer program.

---

[1]OSHA is the acronym for the Occupational Safety and Health Administration, the Federal agency that is charged with monitoring and protecting the well-being of industrial workers.

We can use the list of questions we do have, as well as their answers, to get a handle on the design process and to think about the tasks that are being done when we ask those questions. We usually find that we can *decompose* or break down the process into a sequence of steps (or design tasks) by extracting and naming some of those steps. For example,

When we ask:
- How is the ladder to be used?
- What color(s) should it be?
- How much should it cost?

we are *clarifying the objectives* set for the design by the client.

When we ask:
- Should the ladder be portable?
- How much can it cost?

we are *establishing user requirements* for the design.

When we ask:
- How does OSHA define "safety"?
- Can this ladder be assembled?

we are *identifying constraints* that govern the design.

When we ask:
- Can the ladder lean against a supporting surface?
- Must the ladder support someone carrying something?

we are *establishing functions* for the design.

When we ask:
- How much weight should a safe ladder support?
- What is the "allowable load" on a step?
- How high should someone on the ladder be able to reach?

we are *establishing design specifications* for the design.

When we ask:
- Could the ladder be a stepladder or an extension ladder?
- Could the ladder be made of wood, aluminum, or fiberglass?

we are *generating design alternatives*.

When we ask:
- What is the maximum stress in a step supporting the "design load"?
- How does the bending deflection of a loaded step vary with the material of which the step is made?

we are *modeling* or *analyzing* the design.

When we ask:
- Can the ladder carry the weight, the "design load"?
- Can each step carry its "allowable load"?
- Can someone on the ladder reach the specified height?
- Does the ladder meet OSHA's safety specification?

we are *testing* and *evaluating* the design.

When we ask:
- Is there a more economic design?
- Is there a more efficient design (e.g., less material)?

we are *refining* and *optimizing* the design.

Finally, when we present the client with:
- *fabrication specifications* for the proposed ladder design, and
- the justification for those specifications

we are *documenting* the completed design.

We see quite clearly that the questions we are asking about the ladder design are steps in a process that moves us from an abstract statement of a design objective through increasing levels of detail until we can "build" a model of the ladder, analyze and test that model, optimize and refine some of its features, and then complete the process by documenting both the fabrication specifications and the justification for this particular design. Thus, we are able to identify some of the specific tasks that need to be done in order to complete a design, although it should come as no surprise to state that in order to cover *all* the tasks of engineering design, we might have to identify additional tasks, as well as greater detail for all such tasks.

We will use the balance of this chapter to describe design tasks and processes and how we can begin to organize teams to do the design projects. For now it is worth learning two important lessons from our ladder design project:

- An essential part of an engineering design project is *clarifying* the client's objectives. It is very important for the designer to understand fully what the client wants, and what the users need from the resulting design. This is why good communication between all three parties in the designer–client–user triangle is essential for good design, and why we need to put some thought into eliciting the details of what the client really wants.
- An underlying theme of the remaining tasks in the design process is that of *translating* the client's objectives into the kinds of words, pictures, numbers, rules, properties, etc., that we need to characterize and describe the artifact being designed and its behavior. It should be clear that the tasks of analyzing and modeling, of testing and evaluating, and of refining and optimizing, cannot be done verbally. The documentation of the final design also cannot be completed with words alone. We know from experience that we will also need pictures and numbers, and perhaps other ways

of representing the desired design. The designer must translate from the client's verbal statement into whatever form—or formula—is appropriate to doing the task immediately at hand.

How would we describe the basic function of the ladder? We know that the function the ladder serves is to allow someone to climb up some vertical distance, perhaps to paint a wall, perhaps to rescue a cat from a tree limb. In fact, a design project typically begins with a verbal statement that might state desired features of *function*; of *form* (e.g., an extension ladder or a stepladder); of *intent* (e.g., to be used in a factory or at home); or of legal requirements (e.g., to satisfy government regulations). It is the designer's job to clarify what the client wants and then to translate the client's wishes into more concrete objectives toward which she can work. In the clarification step, the client is asked to be more precise about what is really wanted by asking questions such as: For what purposes is the ladder to be used? Where? How much can the ladder itself weigh? What level of quality do you want in this ladder? How do you define quality? How much are you willing to spend? Note, however, that the degree of precision demanded from the client could change as the design process unfolds.

Some of the questions we asked here in the hope of clarifying the client's wishes obviously connect with a later part of the process in which we begin to make choices, analyze how possibly competing choices interact, assess any trade-offs in these choices, and evaluate the effect of these choices on our overall goal of designing a safe ladder. For example, the form or configuration of the ladder is strongly related to its function: We are more likely to use an extension ladder to rescue a cat from a tree and a stepladder to paint the walls of a room. Similarly, the weight of the ladder certainly has an impact on the efficiency with which it can be used: Aluminum extension ladders have replaced wooden ones largely because they weigh less. The material of which the ladder is made not only influences its weight, but also its cost and its feel: Wooden extension ladders are much stiffer than their aluminum counterparts, so users of the aluminum versions have to get used to feeling a certain amount of "give" or flex in the ladder, especially when it is significantly extended. Thus, a design goal heretofore unmentioned may have emerged: Design a safe, *stiff* ladder.

The next part of the design process is where the "rubber meets the road," for now the client's wishes are translated into a set of *user requirements* that state in greater detail what is wanted from the design by potential users, as well as the client. These user requirements are detailed expressions of the designed object's desired functionality and of its attributes. These user requirements are the basis for elaborating a set of *design specifications* that serve as benchmarks against which the performance of a designed artifact is measured. As we noted in Chapter 1, design specifications are typically stated in one of three ways, depending on the nature of the requirements we want to articulate: prescriptive specifications that specify values for attributes of the designed object, procedural specifications that identify specific procedures for calculating attributes or behavior, and performance specifications that characterize the desired behavior. A design is successful if it meets, or exceeds, the given specifications and satisfies, or exceeds, the client's expectations. We should also expect that the specifications or requirements may evolve or be further refined as a design unfolds.

As a design evolves further, as we futher clarify and flesh out the client's statement, we will come across a vast array of choices to make. At some point in our ladder design, for example, we will have to choose a *type* of ladder, say a stepladder or an extension ladder. Then we will have to figure out how to fasten the steps to the ladder frame. Our choices will be influenced by the desired behavior (e.g., although the ladder itself may flex somewhat, we certainly don't want individual steps to have much give with respect to the ladder frame) and by manufacturing or assembly considerations (e.g., would it be better to nail in the steps of a wooden ladder, use dowels and glue, or nuts and bolts?). Here we are decomposing the ladder into its components or pieces and selecting particular types of components. The translation issue is alive and well here, too, because component choices must be articulated in a "language" naturally conducive to making them. Choosing a particular bolt-and-nut pair to achieve a certain fastening strength requires access to a manufacturer's catalog, as well as to the results of calculations about bearing and shear stresses.

We should also note that as we work through these steps, we are constantly communicating with others about the ladder and its various features. When we question the client about desired properties, for example, or the laboratory director about the evaluation tests, or the manufacturing engineer about the feasibility of making certain parts, we are interpreting aspects of the ladder design in terms of "languages" and parameters that these experts use in their own work. Thus, the design process can't proceed without such interpretations.

We have used this simple example to illustrate how we might *formalize* the design process to make explicit that we are doing some design tasks. We could also say that we are *externalizing* aspects of the process in order to get them from our heads and into some recognizable language(s) for further use. With this in mind, we look now at some descriptions of the design process.

2.2   WAYS OF DESCRIBING THE DESIGN PROCESS

There are many descriptions of the design process, just as there are many definitions of design. One clear starting point might be the set of steps we listed in the last section for the ladder design. Indeed, after some further reflection and with some further augmentation, we will find that a modified version of those steps does pretty well at covering the range of engineering design problems.

2.2.1   A Simple Description of the Design Process

One of the simplest models of the design process highlights three stages. In the first, *generation*, the designer proposes various concepts that are generated, by means unspecified. In the second stage, *evaluation*, the design is tested against the design goals, constraints, and criteria that have been set forth by the client, the user(s), and the designer. In the last stage, *communication*, the design is communicated to the manufacturers or fabricators. A similar three-stage model calls for taking steps to *do research*, to *create*, and then to *implement* a final design, with the meanings of these steps being rather clear by the context. While these two models have the virtue of simplicity, they don't say much at all about what goes on. They are so abstract that they provide little useful advice on how to do a design. To cite one very obvious question, *How* do we generate or create designs?

2.2.2   Describing Phases of the Design Process

In Figure 2.1 we show another, widely accepted model of the design process. This model, too, has three stages, shown in boxes with rounded corners. We also show in this figure the process' starting point and its endpoint in circles. The starting point is the client's statement, which is often identified as the *need* for a design. The endpoint is the final design or the set of fabrication specifications. The first stage of the process depicted in Figure 2.1 is that of *conceptual design*, in which we look for different *concepts* (called *schemes* by some) that can be used to achieve the client's objectives. A concept or scheme is an outline solution to a design problem. Means for achieving each major function have been identified and fixed, as have the spatial and structural relationships of the principal components. Enough details have been worked out so that we can estimate costs, weights, and overall dimensions.

For the safe ladder project, we might have identified the stepladder as our concept. In bridge design, potential conceptual designs might include different types of bridges, say a suspension bridge, a cable-stayed bridge, or an arch—or, if the true goal is to cross a river, tunnels and ferries might be considered as potential conceptual designs. The evaluation of these three concepts or schemes would likely depend on some of the high-level attributes of the client's statement, perhaps including the anticipated span length, financing, type of traffic, and the aesthetic values of the client and the user(s).

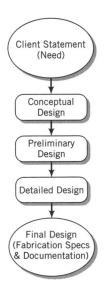

Figure 2.1 A descriptive, three-stage, "linear" model of the design process. This model is very simple in that the process is shown as a linear sequence of objects (*need* and *final design*) connected by three design phases (*conceptual, preliminary,* and *detailed design*).

Interestingly enough, it is the focus of conceptual design on high-level issues, coupled with a relatively weak dependence on technical details, that makes it possible for significant elements of design to be taught, learned, and used early in the professional education of engineers. However, in conceptual design, it is also true that there are technical issues we do not yet address. For example, we cannot predict how subsystems will interact, what subobjectives might conflict, what options may have to be ruled out because of local conditions, and so on, because our design is still a relatively abstract and unrefined concept. Again, for the bridge design, the nature of the gap being spanned (e.g., are we bridging a river or crossing over one or more roadways?) might dictate variations in the lengths of a main span or of subspans, which would in turn affect the configurations of the secondary structural members.

With its focus on tradeoffs between high-level objectives, conceptual design is clearly the most open-ended and abstract part of the design process. The output of the conceptual stage could be one, two, or several competing concepts. Some even argue that conceptual design *should* produce two or more schemes, since early attachment to a single design choice could be a mistake. This tendency is so well known among designers that it has produced a saying: "Don't marry your first design idea." It is also true that at this point in the design process we probably do not have sufficient data to discard concepts or ideas that will later be viewed as "extra" because they did not measure up for one reason or another.

The second phase in this model of the design process is *preliminary design*. (The corresponding European phrase for this stage is the *embodiment of schemes*.) Here the proposed concepts or schemes are "fleshed out," that is, we hang the meat of some preliminary choices upon the abstract bones of the conceptual design. In this phase, we embody or endow design concepts with their most important attributes. We begin to select and size the major subsystems, based on lower-level concerns that take into account the performance specifications and the operating requirements. For the stepladder, for example, we start to size the side rails and the steps, and perhaps decide on the way that the steps are to be fastened to the side rails. For the various bridges, we would lay out the span(s), define the access roads more clearly, estimate the sizes and locations of abutments, and so on. Preliminary design is clearly more technical in nature, and we might use various back-of-the-envelope calculations, as well as rules about size, efficiency, and so on. Here, too, we make extensive use of rules of thumb that reflect the designer's experience. And, in this phase of the design process, we make our final choice from among the proposed concepts.

The third and final stage of this model is *detailed design*. We now turn to refining the choices we made in the preliminary design, articulating our early choices in much greater detail, down to specific part types and dimensions. This phase of design typically follows design procedures that are quite well known. Relevant knowledge is found in design codes (e.g., the ASME Pressure Vessel and Piping Code and the Universal Building Code), handbooks, databases, and catalogs. Design knowledge is often expressed in rather specific rules, as well as in formulas and algorithms. This

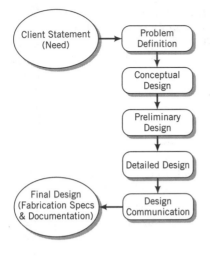

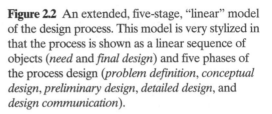

Figure 2.2 An extended, five-stage, "linear" model of the design process. This model is very stylized in that the process is shown as a linear sequence of objects (*need* and *final design*) and five phases of the process design (*problem definition, conceptual design, preliminary design, detailed design*, and *design communication*).

stage of design is typically left to component specialists because the design is now closer to being assembled from a library of standard pieces.

The classic model just outlined can—and should—be formally extended to a five-stage model, as shown in Figure 2.2. We have delineated two stages that are activities that precede and follow, respectively, the sequence of the three-stage model. We call particular attention to these two, "pre-" and "post-processing," design stages because they provide essential transitions between certain artifacts and phases of the design process. *Problem definition*, the pre-processing stage, identifies the work done with the client's statement before conceptual design can begin. *Design communication*, the post-processing stage, identifies the work done after detailed design is complete to present to (and document for) the client and the final design and fabrication specifications.

Although this model of the process is more detailed than the simple ones we discussed in Section 2.2.1, note that we are not much closer to knowing how to do a design. All of the process outlines given thus far in this section are *descriptive*, that is, they describe what is being done rather than suggesting how it ought to be done. We will now present a *prescriptive* process model that suggests or prescribes what ought to done. We obtain this prescriptive model by integrating the design steps or tasks identified in Section 2.1 into the five-stage (descriptive) model we have just presented in Figure 2.2.

2.2.3 Some Prescriptions for the Design Process

We show a *prescriptive* model of the design process in Figure 2.3. It is, as noted, an extension of the five-stage model of Figure 2.2. Our design work begins with the client's statement and ends when the final design is documented to the client. With this in mind, we can situate the ten design tasks identified in Section 2.1 within the five stages (i.e., problem definition, conceptual, preliminary, and detailed design, and design communication) of the design process as follows:

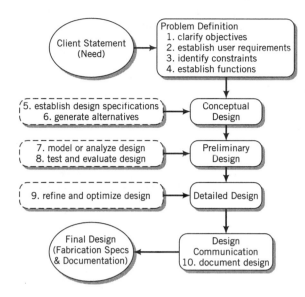

Figure 2.3 A prescriptive, five-stage model of the design process. Like the previous model, this model is also very stylized in that the process is shown as a linear sequence of artifacts (*need* and *final design*) and design phases, within which are situated the design tasks.

1. The *problem definition* phase is devoted to clarifying the objectives set out by the client and gathering the information needed to develop an engineering statement of what the client wants.

   Input: *client's statement*

   Tasks: *clarifying objectives for the design (1)*
            establishing user requirements (2)
            identifying constraints (3)
            establishing functions (4)

   Output: *revised problem statement*
            detailed (weighted) objectives
            constraints
            user requirements
            functions

2. The goal of the *conceptual design* stage of the design process is the generation of concepts or schemes of candidate designs.

   Input: *revised problem statement*
            detailed (weighted) objectives
            constraints
            user requirements
            functions

   Tasks: *establishing design specifications (5)*
            generating design alternatives (6)

   Output: *conceptual design(s)* or *scheme(s)*
            design specifications

3. The goal of the *preliminary design* phase is the identification of the principal attributes of the design concepts or schemes.

Input: *conceptual design(s) or scheme(s)*
              design specifications

Tasks: *model and analyze the conceptual design alternatives (7)*
              test and evaluate the conceptual design alternatives (8)

Output: *a selected design*
              test-and-evaluation results

4. The goal of the *detailed design* phase is the refinement and detailed definition of the final design.

Input: *selected design*
              test-and-evaluation results

Task: *refine and optimize the chosen design alternative (9)*

Output: *proposed fabrication specifications*
              final design review and presentation to client

5. The *design communication* phase is devoted to documenting the fabrication specifications and their justification.

Input: *fabrication specifications*

Task: *document the completed design (10)*

Output: *final report to client containing*
              • fabrication specifications
              • justification for fabrication specifications

We have identified ten design tasks and situated them in the five stages of the design process. For each phase we have shown the *input* to that stage, the *design tasks* to be performed in that stage, and the *output* or product that is produced by that stage. We thus have a "checklist" that we can use to ensure that we have done all of the "required" steps. Similarly, lists like this are also used by design organizations to specify and propagate approaches to design within their groups or firms. However, we should keep in mind that this and other detailed elaborations add to our understanding of the design process only in a limited way. At the heart of the matter is our ability to model the tasks done within each phase of the design process. With this in mind, we will present some means and formal methods for doing these ten design tasks in Section 2.3.

2.2.4   Feedback and Iteration in the Design Process

All of the models we have presented thus far have been "linear" or sequential. In fact, the design process is not linear or sequential. We have intentionally left out two very important elements in design thinking and behavior. The first of these is *feedback*, that is, the process of feeding information about the output of a process (or

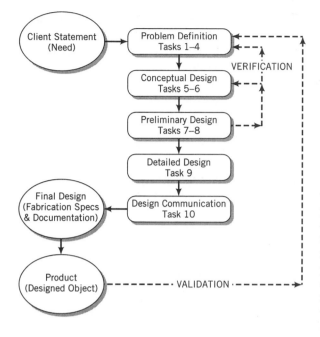

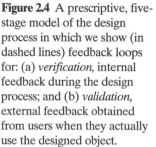

Figure 2.4 A prescriptive, five-stage model of the design process in which we show (in dashed lines) feedback loops for: (a) *verification*, internal feedback during the design process; and (b) *validation*, external feedback obtained from users when they actually use the designed object.

system) back into the process so it can be used to obtain better results from the process. Feedback occurs in two notable ways in the design process, as illustrated in Figure 2.4. The first is an internal *feedback loop* wherein the results of performing the test and evaluation task are fed back into the preliminary design stage to *verify* that the design performs as intended. In addition, and particularly in the context of the emerging practice of concurrent engineering that we detailed in Section 1.3, there will be internal feedback from internal customers, such as manufacturing (e.g., can it be easily made?) and maintenance (e.g., can it be easily fixed?). The second feedback loop is external. It occurs after the final product resulting from a design has been used in the market for which it was intended. User feedback then provides *validation* for the design, assuming it's a successful design!

The second element that we have thus far left out of our process models is *iteration.* We are iterating when we apply a common method or technique at different points in a design process (or in analysis as well), although the repeated applications occur at a different level of abstraction (or on a different scale). These repetitions typically occur at more refined, less abstract points in the design process (and on a finer, more detailed scale in analysis). In terms of the linear, five-stage model depicted in Figure 2.3, we should anticipate repeating the first four tasks in some form in conceptual, preliminary, and detailed design. That is, we always want to keep the original objectives in mind to ensure that we have not strayed from them as we get deeper and deeper into the details of our final design. Of course, this may also mean that we may have reason to do some redesign, in which case we will certainly reiterate tasks 7 (analyzing the design) and 8 (testing and evaluating the design) as well.

Given that there are feedback loops, given that we will repeat or reiterate some tasks, why did we present our process models as linear sequences? The answer is simple. We noted in Chapter 1 that "design is a goal-directed activity, performed by humans." As important as the feedback and iterative elements of design are, it is equally important not to be overly distracted by these adaptive characteristics when learning about—and trying to do—design for the first time. It's also true that, in some sense, feedback loops and the need to repeat some design tasks will occur naturally as a design project unfolds. When we're doing a design for a client, it's only natural to go back and ask the client if the original design statement was properly translated and interpreted. And it's just as natural to show off conceptual designs as they emerge, and to refine these concepts by responding to any resulting feedback and reiterating objectives, requirements, and specifications.

2.2.5 Further Remarks on and Terminology for the Design Process

We have repeatedly said that there is a lot of literature on design, and that there are far more descriptions of the design process than we want to cover here. On the other hand, there are some terms and ideas about design that are often used that are worth adding to our technical vocabulary.

Consider the ladder design problem once again. When we select a type of ladder, say a stepladder or an extension ladder, we are choosing an architecture or a configuration for the ladder. Then, as we noted near the end of Section 2.1, we fastened steps to the ladder frame by choosing nails, glue, or nuts and bolts. This kind of design is called *component selection*, and it occurs after we have decomposed the ladder into its components or pieces, when we are selecting a particular type of component. These kinds of design tasks are thought to be *routine* because the architecture has been specified and the components come from known sources. However, these tasks can be complex if we're designing a product with a lot of parts, say a mainframe computer.

There are cases where the architecture of a device is not specified or known in advance, in which case the task of assembling a set of parts to perform some specified function(s) is more complicated. This class of design problems is gathered under the rubric of *configuration design*, and we see it in cases such as the design of paths along which paper can travel in sophisticated copying machines. The challenge herein derives from obliging pieces of paper to move at very high speeds through a series of photocopying processes in order to have crisp copies emerge rapidly and without wrinkles or tears. The architecture to be designed consists of (1) a path through the machine that avoids obstacles and stops at specified points for very brief periods of time and (2) the set of mechanical components that moves the paper along (and within) that path without jamming. A quick look at a modern copying machine shows just how complex its paper paths can be.

Beyond the task of assembling components lies the more daunting task of *synthesis*, that is, assembling a set of primitive design elements or partial designs into one or more configurations that clearly satisfy a few key objectives and constraints. In fact, synthesis is often considered as the task most emblematic of the design process, so much so that it is often used as the touchstone word to describe design in contrast with analysis. *Analysis* is the task of performing those calculations (or analyses) needed to assess the behavior of the current synthesis or preliminary design. For example, we calculate the bending stress and deflection of a loaded step of the ladder to see how the step behaves under a given set of loads. Notwithstanding this example from ladder design, we should note that analysis is also required early in the design process because we need to think critically or analytically in order to clarify objectives, establish requirements, and elaborate specifications. That is why so much of what follows is devoted to tools and techniques intended to support such critical thinking.

Finally, here, a word or two about creativity. It is often said that synthesis is the creative part of design, leading to a widely-held misperception that creativity is reflected only in synthesis. In fact, to be creative we need analysis and synthesis. We need logical thinking and analysis to decompose a design problem into subproblems, complete the translation of the client's requirements into design specifications, review and revise our project schedule, and prepare a budget for the design process. We need synthesis to prepare conceptual designs or schemes. And creativity is also needed to evaluate alternative conceptual designs and to prepare preliminary designs or prototypes.

2.2.6   On Opportunities and Limits

Most of our effort in this book will be focused on conceptual design, that is, on the first phase of the design process. As a result, we will often be dealing with some very broad themes and approaches in ways that are certainly logical, but not as neat and tidy as a set of mathematical formulas or algorithms. It is ironic that the seeming lack of rigor of the tools we will present for use in conceptual design also makes them very useful more generically as methods for *problem solving*. Indeed, there is something of a debate within the design community as to what constitutes "real design," and as to whether or not such non-rigorous problem solving should be accepted as true engineering design.

As we will show through our examples, the conceptual design tools lend themselves to answering questions that cannot even be properly posed in formal mathematical terms. Further, many design tasks are, in fact, problems that need to be solved. For example, as we will illustrate at length in Chapter 10, the issues surrounding the redesign of the transportation network through and around Charlestown's City Square emerged first as a major transportation *problem*, and then were finally resolved as it became recognized that there were both major transportation and community survival *problems* that needed to be addressed. Some of these tools have been applied to solve other kinds of problems; we can find examples as far afield as selecting "the best" mini-sports utility vehicle and choosing a college (in which case a professor of engineering used one of these tools to help her daughter narrow the field)!

While conceptual design tools can be used in more general problem solving, such applications are fundamentally beyond our concerns. Our interest is in detailing a set of tools and techniques that support conceptual design, a stage from which all engineering design projects originate. Additionally, if these tools prove useful elsewhere, all well and good, as long as our interest in properly formulating and solving conceptual design problems is not lost.

We also note here that we are largely limiting our detailed discussion of design processes to the earliest phases of design. We are not delving into the many processes and activities that must be done before a design can ultimately be turned into a product that a customer can actually buy. While we will discuss some of the related design issues in Chapter 8, we will not be devoting very much space to them here because, as we have said oftentimes already, our primary focus is on successfully doing client-initiated, conceptual design projects.

2.3   STRATEGIES, METHODS, AND MEANS IN THE DESIGN PROCESS

Descriptions of design processes often fail us because they do not tell us *how* to go about the business of generating or creating designs. There are some aspects of the design process where we can be more specific, perhaps even to the point of writing algorithms. Toward this end, we show each of the five stages of the design process in Charts 2.1–5, and for each phase we list: the *input* required to start that phase; possible *sources* of relevant knowledge; the design *tasks* that will be done in this stage; applicable formal design *methods*; some *means* for acquiring and processing additional design and engineering knowledge; and, finally, the *output* or product of that stage. Here we will briefly introduce some of the formal design methods and some of the means of acquiring design-related information methods, as a prelude to the more detailed descriptions in Chapters 3–7. Keep in mind while we are introducing these decision-support techniques and tools that we are explaining *how* we go about designing artifacts, that is, we are describing *thought processes* or *cognitive tasks* that will be done during the design process. We begin with ideas for strategic approaches to design thinking.

2.3.1   Strategic Thinking in the Design Process

A general strategy for thinking about design is called *least commitment*. In other words, don't commit to a particular concept or configuration until forced to by the exhaustion of additional information or of alternate choices. (Remember that saying, "Never marry your first design.") Least commitment is not so much a method as it is a strategy or a good habit of thought. It militates against making decisions before there is a reason to make them. Premature commitments can be dangerous because we might become attached to a bad concept or we might limit ourselves to a suboptimal range of design choices. Least commitment is of particular importance in conceptual design because the consequences of any one design decision are likely to be propagated far down the line.

Another important strategy of design thinking is to apply the power of *decomposition*, that is, of decomposing or breaking down larger problems (or entities or ideas) into smaller subproblems (or subentities or subideas). These smaller subproblems are usually smaller and easier to solve or otherwise handle.

Chart 2.1 PROBLEM DEFINITION: *input*, knowledge *sources*, design *tasks*, formal design *methods*, *means* for processing knowledge, and *output*

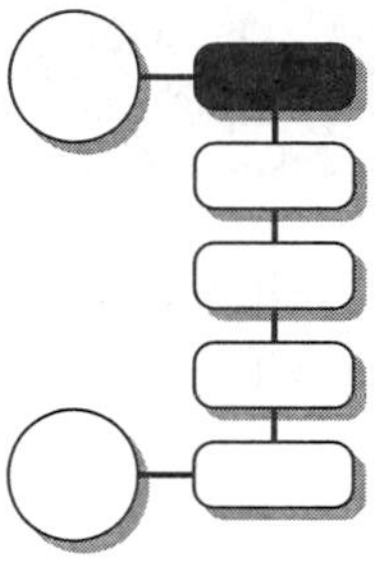

1. The *problem definition* phase is devoted to clarifying the objectives set out by the client and gathering the information needed to develop an
   engineering statement of what the client wants.
                    Shelby

Input:	*client's statement*
Sources:	*literature on state-of-the-art*
	experts
	codes and regulations
Tasks:	*clarifying design objectives (1)*
	establishing user requirements (2)
	identifying constraints (3)
	establishing functions (4)
Methods:	*objectives tree*
	pairwise comparison chart
	weighted objectives tree
	function means tree
	functional analysis
	requirements matrix
Means:	*literature review*
	brainstorming
	user surveys and questionnaires
	structured interviews
Output:	*revised problem statement*
	detailed (weighted) objectives
	constraints
	user requirements
	functions

That is why decomposition is often labelled as "divide and conquer." We have to keep in mind that subproblems can interact, so we must ensure that they do not violate the assumptions or constraints of complementary subproblems.

2.3.2 Some Formal Methods for the Design Process

We now present brief introductions to the formal design methods that are listed in Charts 2.1–5 for the five stages of the design process.

We build objectives trees in order to clarify and better understand a client's project statement. Objectives trees are hierarchical lists that branch out into

Chart 2.2 CONCEPTUAL DESIGN: *input*, knowledge *sources*, design *tasks*, formal design *methods*, *means* for processing knowledge, and *output*

2. The goal of the *conceptual design* stage of the design process is the generation of concepts or schemes of candidate designs.

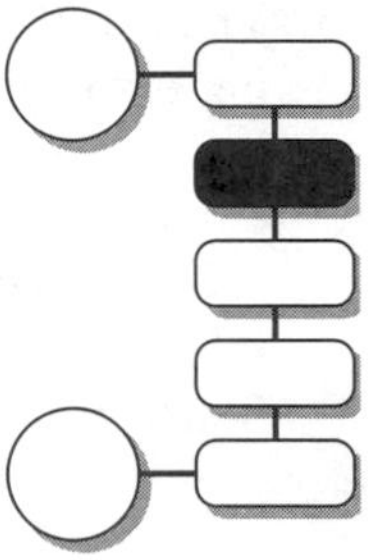

    Kyle

Input:	*revised problem statement*
	detailed (weighted) objectives
	constraints
	user requirements
	functions
Sources:	*competitive products*
Tasks:	*establishing design specs (5)*
	generating design alternatives (6)
Methods:	*performance specification method*
	quality function deployment (QFD)
	morphological chart
Means:	*brainstorming*
	synectics and analogies
	benchmarking
	reverse engineering (dissection)
Output:	*conceptual design(s)* or *scheme(s)*
	design specifications

tree-like structures and in which the objectives that designs must serve are clustered by subobjectives and then ordered by degrees of further detail. Thus, starting at the highest level of abstraction of an objectives tree, we have a top-level design goal, which is the client's project statement. We move down the list or tree to more detailed, more specific subobjectives. In Section 3.1 we detail how we construct objectives trees and what kind of information we learn from them.

We rank design objectives in two ways. The first is the pairwise comparison chart, a relatively simple device in which we list the objectives as both rows and columns in a matrix or chart and then compare them on a pair-by-pair basis, proceeding in a row-by-row fashion. The comparison chart is useful for ranking objectives early in the design process, and it is helpful whenever we must choose among competing functions, attributes, or requirements. The second ordering tool is the *weighted objectives tree*, in which subobjectives are given weights that are normalized with respect to the objective immediately above them in the tree and with respect to other subobjectives at the same level in the tree. These weightings or rankings are sometimes considered more "reliable" because they are systematically normalized. We describe pairwise comparison charts and weighted objectives trees in Section 3.3.

Chart 2.3 PRELIMINARY DESIGN: *input*, knowledge *sources*, design *tasks*, formal design *methods*, *means* for processing knowledge, and *output*

3. The goal of the *preliminary design* phase is the identification of the principal attributes of the design concepts or schemes.

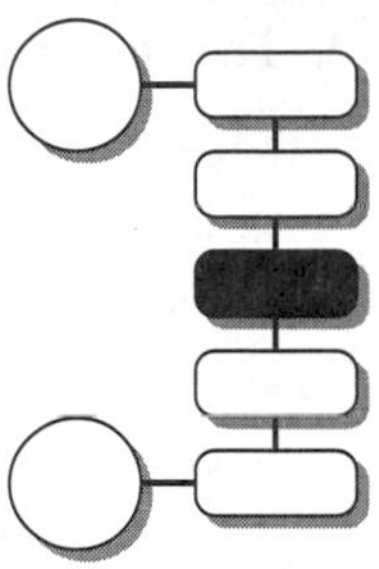

> Jeff

Input:	*conceptual design(s)* or *scheme(s)*
	design specifications
Sources:	*heuristics (rules of thumb)*
	simple models
	known physical relationships
Tasks:	*model, analyze conceptual designs (7)*
	test, evaluate conceptual designs (8)
Methods:	*weighted objectives tree*
	pairwise comparison chart
Means:	*metrics definition*
	laboratory experiments
	prototype development
	simulation and computer analysis
	proof-of-concept testing
Output:	*a selected design*
	test-and-evaluation results

Functional analysis is helpful for focusing on what a design must achieve by identifying and listing in an organized way its inputs and its outputs. The starting point for analyzing the functionality of a proposed device is usually a "black box" with a clearly delineated boundary between the device and its surroundings (recall Simon's definition of design in Section 1.3). In functional analysis, we decompose overall function(s) into subfunction(s). This is achieved by tracking the flow of materials, signals, etc., through the device and detailing the material- or signal-processing needed to produce the desired functions. We present functional analysis and the related *function-means tree*, which is also a relative of the objectives tree, in Section 5.1.

The *requirements matrix* is a matrix or chart in which we attempt to match up the functions and attributes of a device wanted by a client and/or users with the engineering properties of those functions and attributes. In other words, we identify the technical or engineering implications of wanting a device to behave in a certain way.

The *performance specification method* provides support for the elaboration of the design specifications that are the designer's target for a design project. The aim is to list solution-independent attributes and performance specifications (i.e., "hard numbers") for both required and desired features of a design concept. We describe performance specifications and their use in Section 5.2.

Chart 2.4 DETAILED DESIGN: *input*, knowledge *sources*, design *tasks*, formal design *methods*, *means* for processing knowledge, and *output*

4. The goal of the *detailed design* phase is the refinement and detailed definition of the final design.

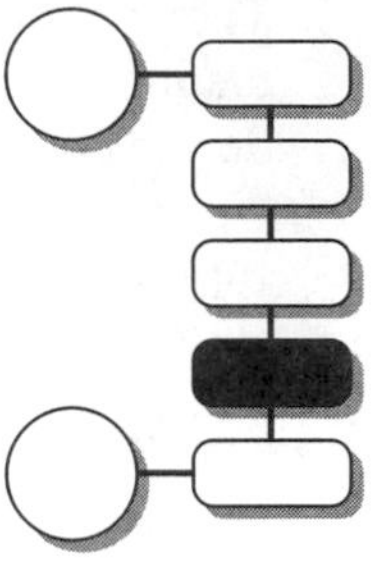

   CHAP

   Input: *selected design*
              test-and-evaluation results

   Sources: *design codes*
              handbooks
              local laws and regulations
              suppliers' component specifications

   Task: *refine, optimize the chosen design (9)*

   Methods: *discipline-specific CADD*

   Means: *formal design reviews*
              public hearings (if applicable)
              beta testing

   Output: *proposed fabrication specifications*
              final design review for client

Chart 2.5 DESIGN COMMUNICATION: *input*, knowledge *sources*, design *tasks*, formal design *methods*, *means* for processing knowledge, and *output*

5. The *design communication* phase is devoted to
   documenting the fabrication specifications and
   their justification.

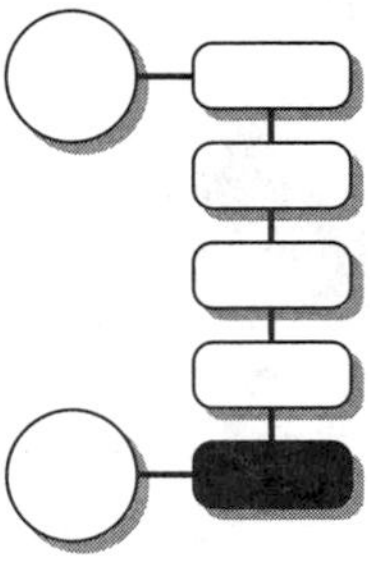

   Michael

   Input: *fabrication specifications*

   Sources: *feedback from clients and users*
              required deliverables

   Task: *document the completed design (10)*

   Methods: *N/A*

   Means: *N/A*

   Output: *final report to client containing*
              (1) fabrication specifications
              (2) justification for fabrication specs

(A more advanced tool, *quality function deployment*, commonly known as QFD, builds on the performance specification method with the goal of achiev-ing higher-quality products. Used in large-scale manufacturing, QFD calls for charting client and user requirements and engineering attributes in a format that makes it possible to relate each to, and to weigh each against, one another. The main point is to erect a *house of quality* that exposes both positive and negative interactions of the engineering specifications, thus enabling a designer to anticipate and weed out performance conflicts. We will not detail QFD here.)

The *morphological chart*, sometimes called a *function–means table*, is used to identify the ways or means that can be used to make the required function(s) happen. A "morph" chart provides a framework for visualizing a *design space* for generating and exploring alternative designs, that is, an imaginary "plane," "room," or "space" that we can use to collect, identify, store, and study all of the potential designs that might solve our design problem. We describe design spaces more formally in Section 6.1 and morph charts in Section 6.3.

2.3.3   Some Means for Acquiring and Processing Design Knowledge

Here we describe the means, listed in Charts 2.1–5, by which information is gathered and analyzed for ultimate use in the formal design methods. These ways or means are tools that have been developed for a number of disciplines, and we can organize them roughly into three categories: means for obtaining information; means for analyzing the information obtained and testing outcomes against desired results; and means for obtaining feedback from clients, users, and stakeholders (i.e., other interested parties). In many cases, the means are so widely applied we view their detailed description as being beyond our scope. In some cases, however, the means are so important to design teams that we will provide expanded discussions later in the book.

2.3.3.1   *Means for Acquiring Information (Including Design Alternatives)*

The classic method of determining the state of the art and prior work in the field, the *literature review*, is so well documented and understood that it might seem unnecessary for us to comment on it. However, is is worth noting that the relevant literature can be both vast and greatly dependent on the stage or phase of the design. In the preprocessing and conceptual stages we need literature reviews to enhance our understanding the nature of potential users, the client, and the design problem itself. We are then likely to turn to prior or existing solutions, including product advertising and vendor literature. In preliminary design, we are likely to look more toward the technical literature regarding the physical properties of possible solutions concepts. In the detailed design stage, we will want to look at handbooks, compendia of material properties, design and legal codes, etc.

User surveys and questionnaires are used to do market research. They naturally tend to focus on identifying both user understanding of the problem space and user response to possible solutions. Market research could be very important to the designer in the early stages of a design project, to help clarify and better understand the problem itself, so that questions are necessarily open-ended. Later surveys may be used together with pairwise comparison and morphological charts to determine appropriate selection criteria and weights.

Focus groups are an expensive way of allowing a design team to observe the response of appropriately selected users and others to potential designs. Since the intelligent use of focus groups demands considerable sophistication in matters psychological, they are not generally used by student design teams.

On the other hand, *informal interviews* are often undertaken very early in a design project, when the team is still trying to define the problem sufficiently to plan an approach. While informal interviews are relatively simple to conduct, it is important to be sensitive to the time and other constraints of the interviewee. All too often a design team will simply show up and ask seemingly random questions, with the dual effect of highlighting their ignorance while doing little or nothing to dispel it. There are ways to reduce this problem, including sending the interviewee copies of the topics and questions in advance and undertaking extensive literature research prior to conducting interviews.

One way to get information from users or *experts* is to combine elements of the survey or questionnaire with the flexibility possible in informal interviews in a *structured interview*. Here the interviewer has a set of questions to ask of an interviewee, and they may or may not be made available to the interviewees. In addition to using the question set to get direct answers, the interviewer is afforded the opportunity to expand upon a particular question and to open up new areas. The use of a structured set of questions also assures the person being interviewed that the interview has both purpose and focus, and it ensures that interesting side issues do not prevent the key matters from being covered.

One way that design teams can acquire further insight is to do some *brainstorming*, an activity that allows the participants to generate related, or perhaps unrelated, ideas that are listed but not evaluated until a later time. The freewheeling nature of brainstorming can be very helpful in opening up new avenues for research and analysis. However, as we will note in our more detailed discussion of brainstorming in Section 6.2, it is very important that team members maintain a high level of respect for the ideas of others and that all ideas are captured or listed as they are offered.

Design teams can draw upon their own abilities to discover and explore relationships and similarities between ideas and solutions that initially seem unrelated. In an environment free of criticism and evaluation, similar to that in brainstorming, the team conducts a *synectic* activity in which it tries to uncover or develop *analogies* between one type of problem and other types of problems or phenomena. For example, a team might seek to redefine a problem into terms of some other matter solved by an earlier team. Or, the team might initially seek to find the most outrageous solutions to a problem and then look for ways that such solutions could be made useful. We will say more about synectics in Section 6.2, but we note that it is time consuming and demands serious commitment on the part of the design team. Consequently, synectics is generally more widely used in industrial design settings than in academia.

Finally, design teams often look at "what's out there" when developing new products through two related activities. In one case, *competitive products* are *benchmarked*, that is, designers literally look at similar products that are already available in the marketplace and try to evaulate how well those products perform certain functions or exhibit certain features. These competitive products then serve as standards against which proposed design solutions can be compared, in the spirit of "building a better mousetrap" by knowing how well existing mousetraps perform. The second activity, called *reverse engineering* or *dissection*, consists of examining competitive or similar or prior products in great detail by dissecting them or literally taking them apart. The idea here is to determine why a given product or device was designed in the way it was, with the intention of finding better ways of performing the same or similar subfunctions.

2.3.3.2   *Means for Analyzing Information and Testing Outcomes*

An important first step in determining whether or not concepts might work as solutions to problems is to *define the metrics* against which an outcome can be measured. Of particular interest is the relationship between the expression of a design's stated objectives and constraints in the design specifications, on the one hand, and the statement of acceptable and desirable design outcomes, on the other. This is an extremely important topic that we will discuss in depth in Section 5.3.

In some cases, we find it possible to gather and assess data and information about a potential design solution by undertaking *laboratory experiments*. For example, if the solution involves a structure, it may be possible to measure the stress/strength relationships of critical parts of the design in a laboratory test.

One very important means of determining whether or not a design can perform its required functions is *prototype development*. Here a prototype or test unit may have some or all of the functional characteristics of the final design, even though it may not look at all like the expected end product. In fact, it is often the case that early prototypes have only a small subset of the required functionality, both to shorten development time and to reduce costs. Prototypes may also be instrumented to support laboratory or other tests. In many cases, such as large-scale civil engineering projects, development of prototypes may not be feasible or meaningful.

A crucial step along the path from conceptual design to detailed design is *proof-of-concept testing*. Such testing involves establishing a formal means of determining whether or not the concept under consideration can reasonably be expected to meet the design requirements. In many cases, a proof-of-concept test will not require that the tested unit survive or even perform the stated function(s). Instead, the key issue is to show that a design can be modified to fulfill the functions under certain prespecified conditions. And, as with any scientifically based test, it is important that the team determine in advance what outcome would be sufficient to accept or reject the concept.

In many cases, we can't develop or test a prototype, perhaps because of cost, size, hazard, or some other reason. In such cases, we often resort to *simulation*, wherein we exercise a computer model or a physical model of a proposed design to simulate its performance under a stated set of conditions. This presumes that the device being modeled, the conditions under which it operates, and the effects of the operations are sufficiently well understood to make the results useful. One outstanding example of such simulation is the use of wind tunnels and related computer analyses to assess the effects of wind loading on tall buildings and on long, slender suspension bridges.

Closely related to simulation, *computer analysis* involves the development of a computer-based model, which may consist simply of the equations relevant to describing the design, and application of various analytic, discipline-based techniques. These include finite element analysis, integrated circuit modeling, failure mode analysis, criticality analysis, etc. Computer-based models are used widely in all engineering disciplines and they become even more important as a design project moves into detailed design. That's why we refer to such tools as *discipline-specific CADD* systems in Chart 2.4, or, discipline-specific computer-aided design and drafting systems.

2.3.3.3   Means for Generating Feedback

Among the most important means for obtaining feedback from both clients and users *are regularly scheduled meetings with clients and users* at which the progress of the design project, including articulation of the various stages of the design process, is tracked and discussed. We assume throughout our discussions that the design team is always communicating with both clients and users. We will often suggest that various formal design results be reviewed with them.

At certain stages of the design process it is a standard practice to hold a *formal design review* wherein the current design is described to the client(s), selected users, and/or other stakeholders. These meetings typically include sufficient technical detail that the implications of the design can be fairly explored and assessed. It is particularly important that young designers become comfortable with the "give and take" that often accompanies such reviews. While it may seem harsh to be asked to justify various technical details to clients and others, it is usually beneficial because implicit unwarranted assumptions and errors or oversights are uncovered.

In some design environments, relevant laws or public policies require that *public hearings* be held for the purpose of exposing the design to public review and comment. While it is beyond our scope to consider such hearings in detail, it is useful for designers to understand that just as teams are increasingly the internal organizational structure of choice for design, public hearings and meetings are increasingly the norm for major design projects, even when the client is a private corporation.

We have already noted that *focus groups* are important sources of user input on problem definition. Such groups are also widely used to assess user reaction to designs as they near adoption and marketing.

In some industries, most notably software, it is the practice to release an almost-but-not-quite-finished version of a product to a limited number of users. This practice, called *beta testing*, allows the designers to receive feedback about their product before it reaches a larger market. Beta testing, which also allows errors in design or implementation to be uncovered, will not be discussed further in this text.

2.4   GETTING STARTED: MANAGING THE DESIGN PROCESS

Just as there are many models used to describe the design process, there are also many different approaches to describing project management. At this stage, it will be useful for us to provide a brief road map of the project management path to be followed in a design project and to introduce several key points that are important immediately in setting up a project and a project team. In Chapter 4, we will examine project management tools in more depth and we will show how some of them can be useful even for small-scale projects.

Our road map of the management of a design project is given in Figure 2.5, in direct analog with Figure 2.3 for the design process. This figure highlights the fact that project management follows a path from an initial understanding of the problem and an associated project (project definition), to developing and applying a plan to solve the problem (project framework), to organizing that plan in light of time and other resource constraints (project scheduling), and then, as the project unfolds, keeping track of time, work, and cost (project tracking, evaluation, and control).

In Chapter 4 we will examine a number of tools that will prove useful along this path. At this time, however, it is worthwhile to note how this framework affects the project team at the very beginning. Many of the activities associated with project definition, such as the feasibility study, the orientation meeting, and setting the overall schedule and budget, are often beyond the control of the design team. Nevertheless, the team will usually devote its initial meetings and activities to trying to better understand these issues. We will look formally at some of the design-specific tools for problem definition in the next chapter. However, one activity that cannot be put off until later is the organization and development of the project team. We turn to that now.

2.4.1 Organizing Design Teams

Design is an activity that is increasingly done by teams, rather than by individuals acting alone. New products, for example, are often developed by teams that include designers, manufacturing engineers, and marketing experts. These teams

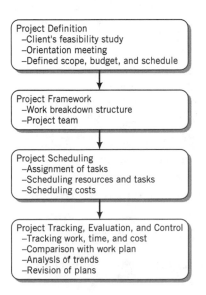

Figure 2.5 Managing a design project follows an orderly process, beginning with the client's understanding of the problem. At the early stages, the design team is concerned with understanding the problem and making plans to solve it. Later, the focus shifts toward project control and staying on plan. Adapted from (Orberlander 1993).

are put together to garner the diverse skills, experiences, and viewpoints needed to design, manufacture, and sell new products successfully. Even in fields where we might expect most of the designers to share similar backgrounds or professional credentials, we can expect to find teams performing a large share of design activity, if only to ensure effective communication across complex product-related disciplines. This dependence on teams is not surprising if we reflect on the stages, methods, and means we have been discussing. Many of the activities we have discussed have a strong dependence on understanding the views and perspectives of others. Many of the methods document the outcomes of applying different talents and skills to understanding the needs of others. Consider, for example, the difference between laboratory testing and computer-based analysis of a structure. While both require technical knowledge of structural mechanics, the separate laboratory and computer skills require years of investment to acquire and master the specific testing and analysis techniques involved. Thus, there may be considerable value in constructing a team whose members have all the needed skills and then depending upon them to work together successfully. In this section we briefly introduce some aspects of team formation and performance, and we then relate them to one of the means for generating ideas discussed above, namely brainstorming.

2.4.1.1   Stages of Group Formation

Groups and teams are such an important element of human enterprise that we should not be surprised to learn that they have been extensively studied and modeled. One of the most useful models of group formation suggests that almost all groups typically undergo five stages of development. These stages have been rather memorably named as the forming, storming, norming, performing, and adjourning stages. We will use this five-stage model to describe some of the elements of group dynamics that are often encountered in engineering design projects.

When we are initially assigned to a team or group, most of us experience a number of feelings simultaneously. These feelings range from excitement and anticipation to anxiety and concern. We may worry about our ability—or that of our teammates—to perform the tasks asked of us. We may be concerned about who will show the leadership needed to accomplish the job. We may be so eager to get started that we rush into assignments and activities before we are really ready to begin. Each of these feelings and concerns are elements of the *forming* stage of group development. The forming stage has been characterized by a number of aspects and behaviors including:

- becoming oriented to the task at hand—in our case, design;
- becoming acquainted with the other members of the team;
- testing group behaviors in an attempt to determine if there are common viewpoints and values;
- becoming dependent upon whoever is believed to be "in charge" of the project or task; and
- attempting to define some initial ground rules, usually by reference to explicitly stated or externally imposed rules.

In this stage, the team members may often do or say things that reflect their uncertainties and anxieties. It is important to recognize this because judgments made in the forming stage may not prove to be valid over the life of the project.

After the initial or forming stage, most groups come to understand that they will have to take an active role in defining the project and the tasks needed to complete it. At this time the group may go through a period of resistance—even resentment—to the assignment, and it may challenge established roles and norms. This period of group development is known as the *storming* phase. It is often marked by intense conflict as team members decide for themselves where the leadership and power of the team will lie, and what roles they must individually play. At the same time, the team will usually be redefining the project and tasks, and discussing opinions about the directions the team should explore. Some characteristics of the storming phase are:

- resistance to task demands,
- interpersonal conflict,
- venting of disagreement, often without apparent resolution, and
- struggle for group leadership.

For the design team, the storming phase is particularly important because there is often already a high level of uncertainty and ambiguity about client and user needs. Some team members may want to rush to solutions, and they may consider a more thoughtful exploration of the design space simply as stubbornness. At the same time, most design teams will not have as clear a leadership structure as, for example, a construction, manufacturing, or a research project. For these reasons, it is important for effective teams to recognize when the team is spending too long in the storming phase and to encourage all team members to move to the next phases, norming and performing.

At some point, most groups do agree on a method of working together and on a set of acceptable behaviors, or norms, for the group. This important period in the group's formation defines whether, for example, the group will insist that all members attend meetings, whether insulting or other disrespectful remarks will be tolerated, and whether or not individuals will be held to high or low standards for acceptable work. It is particularly important that team members understand and agree to the outcome of this so-called *norming* phase, because it may well determine both the tone and the quality of subsequent work. Some characteristics of the norming phase include:

- clarification of roles in the group,
- emergence of informal leadership,
- development of a consensus on group behaviors and norms, and
- emergence of a consensus on the group's activities and purpose.

Significantly, norming is often the stage at which members decide just how seriously they are going to take the project. As such, it is important for team members who want a successful outcome to recognize that simply ignoring unacceptable behavior or poor work products will not be productive. For many teams, the norms that are established at the norming stage become the basis for behavior for the remainder of the project.

After the team has passed through the forming, storming, and norming stages, the team will, we hope, reach the stage of actively doing their project. This stage, called the *performing* phase, is the one that most teams hope to reach. At this point, team members are focusing their energy on the tasks themselves, conducting themselves in accordance with the established norms of the group, and generating useful solutions to the problems they face. The characteristics of the performing phase include:

- clearly understood roles and tasks,
- well-defined norms that support the overall goals of the project,
- sufficient interest and energy to accomplish tasks, and
- emerging solutions and results.

This is the point in team development at which it becomes possible for the goals of the team to be fully realized.

The final phase that teams typically pass through is referred to as *adjourning*. This stage is reached when the group has accomplished its tasks and is preparing to disband. Depending on the extent to which the group has forged its own identity, this stage may be marked by members feeling regret that they will no longer be working together, and some team members may act out some of these concerns in ways that are not consistent with the group's prior norms. These feelings of regret typically emerge after design or other groups have been working together for a very long time, much longer than one or two academic semesters, because it usually takes a lot of time for such complete group identity to develop.

One point to note about these stages of group formation is that teams will typically pass through each of them *at least once*. If the team undertakes significant changes in composition or structure, such as a change in membership, or a change in team leadership, it is likely that the team will revisit the storming and norming phases again.

2.4.1.2   *Team Dynamics and Brainstorming*

In Section 2.3.3.1 we discussed a number of means of gaining information, generating and evaluating ideas, and obtaining feedback on outcomes. Several of these means were based on putting the team and other interested parties into situations that would encourage a free flow of ideas. Brainstorming and synectics in particular are based on the notion that one person's ideas may serve to stimulate other team members to come up with better alternatives. In this section, we briefly summarize brainstorming and relate it to our discussion of the stages of group formation. We will see that our warnings to defer solutions until after the problem is sufficiently understood are not only consistent with our models of the design process, but also with how teams can function best.

Brainstorming is a classic technique for generating ideas and solutions to problems. In essence, brainstorming consists of the members of a group each offering an idea one after another without any evaluation of the idea. One or more members of the team act as "scribe," writing down each idea offered for later discussion and review. Typically, a team will form a circle or sit around a table and, after a brief review of the problem for which ideas are being sought, offer ideas one after another. Each member of the group should offer an idea when his or her turn comes, even if the idea is poorly formed or even silly. At some point, it becomes permissible for a team member to pass, but this must be done explicitly. One of the hoped-for outcomes of such brainstorming is an exhaustive list of possible solutions to the problem. Another result we hope for is that one member's ideas that may not be practical at first may stimulate the creativity of another member who then leverages the first idea into a more useful one. It is extremely important that the participants understand the need to *separate* the generation of ideas from their evaluation. Brainstorming is a technique for generating ideas which, by its design and nature, is likely to lead to many ideas that will subsequently be rejected. At its core, brainstorming is an activity based on respect for the ideas of others, even to the point of being willing to suspend judgment on them temporarily. If a team chooses to focus on evaluating ideas on the spot, it is likely to limit the willingness of team members to offer ideas. The team thus limits the extent to which later changes or "piggybacked" ideas may make earlier offerings possible.

Our previous discussion of group formation stages becomes relevant in helping us to understand when it is likely to be effective to engage in activities such as brainstorming. Clearly in the forming and storming phases, trust and confidence are likely to be absent from the team's dynamics. Indeed, the team may still be trying to define what the real task of the team is. The team is not likely to be in agreement about how seriously it will attempt to meet the team's nominal goals. As such, the team is almost certainly going to be unable to undertake effective brainstorming just yet. On the other hand, at the norming stage, the team is likely to be developing a consensus about norms of behavior, and this may make it possible for the team to set the sort of respect-based rules that brainstorming requires. The team is most able to engage in brainstorming during the performing stage. This implies that models of design that allow for considerable research and problem definition at the early stages are most likely to be consistent with the underlying dynamics of how the team will perform.

2.4.2 Constructive Conflict: Enjoying a Good Fight

Whenever people get together to accomplish tasks, conflict is an inevitable by-product. Much of this conflict is healthy, a necessary part of exchanging ideas, comparing alternatives, and resolving differences of opinion. Conflict can, however, be unpleasant and unhealthy to a group, and it can result in some team members feeling shut out or unwanted by the rest of the group. Thus, a solid understanding of the notions of constructive and destructive conflict is an essential starting point for team-based projects. Even in those cases where team members have been exposed to conflict management skills and tools, it is useful to review them at the start of every project.

The notion of constructive conflict had its origins in research on management conducted in the 1920s. It was observed that the essential element underlying all conflict is a set of *differences*: differences of opinion, differences of interests, differences of underlying desires, etc. Conflict is unavoidable in interpersonal settings, so it should be understood and used to increase the effectiveness of all of the people involved. To be useful, however, conflict must be constructive. *Constructive conflict* is usually based in the realm of ideas or values. On the other hand, *destructive conflict* is usually based on the personalities of the people involved. If we were to list situations where conflict is useful or healthy, we might find such items as "generating new ideas" or "exposing alternative viewpoints." A similar list of situations in which conflict reduces a team's effectiveness would probably include items such as "hurting feelings" or "reducing respect for others." This difference between destructive (i.e., personality-based) conflict and constructive (i.e., idea-based) conflict must be recognized by a team from the outset. While a team is establishing norms, and even before these have been formalized or agreed to in the "norming" phase, the team must establish some basic ground rules that prohibit destructive conflict, and it must enforce them by responding to violations of the ground rules. Destructive conflict, including insults, personally denigrating remarks, and other such behaviors, must be stopped from the outset or it will become part of the team's culture.

Once we note this difference between constructive and destructive conflict, it is useful to recognize various ways that persons can react to conflict. Five basic strategies for resolving conflicts have been identified

- *avoidance*—ignoring the conflict and hoping it will go away;
- *smoothing*—allowing the desires of the other party to win out in order to avoid the conflict;
- *forcing*—imposing a solution on the other party;
- *compromise*—attempting to meet the other party "halfway"; and
- *constructive engagement*—determining the underlying desires of all the parties and then seeking ways to realize them.

The first three of these (avoidance, smoothing, and forcing), all turn on the notion of somehow making the conflict "go away." Avoidance rarely works, and serves to undercut the other party's respect for the person who is hiding from the conflict. Smoothing may be appropriate for matters where one or both of the parties in conflict really don't care about the issue at hand, but it will not work if the dispute is over a serious or important matter. Once again, the respect of the person "giving in" may become lost over time. Forcing is only likely to be effective if the power relationships are clear, such as in a "boss-subordinate" situation, and even then, the effects on morale and future participation may be very negative. Compromise, which is a first choice for many people, is actually a very risky strategy for teams and groups. At its core, it assumes that the dispute is over the "amount" or "degree" of something, rather than on a true underlying principle or difference. While this may work in cases such as labor rates or times allocated for something, it is not likely to be effective in matters such as choosing between two competing design alternatives. (We cannot, for example, compromise between a tunnel and bridge by building a suspension tunnel.) Even in those cases where compromise is possible, we should expect that the conflict will be likely to reoccur after some period of time. Labor and management, for example, often compromise on wage rates—only to find themselves revisiting the very same ground as soon as the next contract opens up. That leaves us with constructive conflict as the only tool that holds the possibility of stable solutions to important conflicts.

Constructive conflict takes as its point of departure an honest telling and listening to each party's underlying desire. Each side must reflect on what it truly wants from the conflict, and then honestly report that to the other parties. They must also listen carefully to what the other party really seeks. In many cases, the conflict is not based on the apparent problem, but rather because each party's underlying desires are different. One paradigmatic anecdote tells of a case where the originator of the idea of constructive conflict, Mary Parker Follett, was working in a library at Harvard on a wintry day, with the windows closed. Another person came into the room and immediately opened one of the windows, thus setting the stage for a conflict and for identifying a way to resolve that conflict. Each of the five alternatives outlined above was nominally available, but most of them were clearly unacceptable. Smoothing, for example, by doing nothing, would have left Ms. Follett cold and uncomfortable. Compromise, opening the window halfway, did not appear to be a meaningful alternative. Instead, she chose to speak with the other person and express her desire to keep the window closed in order to avoid the chill and draft. The other party agreed that this was a good thing, but noted that the room was very stuffy, which in turn bothered his sinuses. Both agreed to look for a reasonable solution to their underlying desires. They were fortunate to find that an adjacent work area also had windows that could be opened, thus allowing for fresh air to enter indirectly without creating a draft. Obviously, this solution was only possible because the configuration of the library allowed it. Nevertheless, they would not have even looked for this outcome except for their willingness to discuss their underlying desires. There are many cases where this will not work, such as when two persons wish to each marry the same third person. There are, however, many cases where constructive engagement will work, both to increase the solution space available to the parties in conflict and to heighten the understanding and respect of the other party. Even when the team is forced to revert to one of the "win-lose" strategies, the team should always consider constructive engagement for resolving important conflicts.

2.5   ILLUSTRATIVE EXAMPLES: INTRODUCTIONS AND DESCRIPTIONS

In Section 1.5 we promised the elaboration of three design examples that illustrate the kinds of work that engineers do when they are producing concepts or ideas to solve a given design problem. The first of these projects, the design of a container for a new juice product, will be used as the design project for which we introduce and explain formal design methods. A summary of that design project is as follows:

1. Design a container to deliver a children's beverage
   Designers: Clive L. Dym and Patrick Little
   Clients: Great American Food and Tobacco (GRAFT) and
                         Bringing Juice Into Children (BJIC)
   Users: Children living both in the United States and abroad
   Project statement: Design a bottle for our new children's juice product.

The second design project that we illustrate is based on the work done by students in a freshman design course at Harvey Mudd College. We use their results, with both their permission and some post-project critiques of our own, to further explain how formal design methods are used. These further explanations appear at the ends of the chapters where particular formal design methods are introduced. The particular project is:

2. Design a chicken coop to be built and used by a Mayan cooperative in Guatemala
   Designers: Teams of Harvey Mudd College freshmen in a design
                         class
   Client: Xela-Aid, a humanitarian aid group
   Users: Mayans living in the town of San Martin Chiquito, in
                         the rural province of Quetzaltenango, Guatemala
   Project statement: Design and produce a chicken coop to be used by several families in a Guatemalan village. The coop should be made of indigenous materials and its design should take into account the climate and environment of a Central American rain forest.

The last design project that we illustrate is based on work done over a fifty-year span to improve a transportation network and hub significantly, as well as the urban environment in which they are placed. The City Square project was intended to enable automobile and rail commuter traffic between Boston and its northern and western suburbs, through Charlestown, Massachusetts. This conceptual design problem clearly illustrates the many factors that go into large-scale engineering projects in their early stages, when choices are being made between tunnels or bridges. Among the design concerns are cost, implications for future expansion, and preservation of the character, environment, and even the view of the affected neighborhoods. This project is a wonderful example of how conceptual design thinking can significantly influence some very "real world" events. The particular project, collected as a unified case study in Chapter 10, is:

3. Design the transportation network and hub through City Square in Charlestown, Massachusetts
   Designers: Various companies and agencies over several decades

Client: The Commonwealth of Massachusetts

Users: Drivers who want to get in and out of downtown Boston

Project statement: Enable automobile and rail commuter traffic between Boston and its northern and western suburbs, through Charlestown.

2.6 NOTES

Section 2.1: The stepladder example derives from the freshman design course taught at Harvey Mudd College and is briefly described in (Dym 1994b).

Section 2.2: As with definitions of design, there are many descriptions of the design process and many of them can be found in (Cross 1989), (Dym 1994a), (French 1985, 1992), (Pahl and Beitz 1984), and (VDI 1987). Further descriptions of the tasks of design can be found in (Asimow 1962), (Dym and Levitt 1991a), and (Jones 1981). Component selection and configuration design are described in (Darr and Dym 1998). A knowledge-based design system that addressed the copier paper path problem is decribed in (Mittal, Dym, and Morjaria 1986) and summarized in (Dym and Levitt 1991a). Applications of design processes to design-for-manufacturing are given in (Dixon and Poli 1995). Examples of the application of conceptual design tools as problem-solving tools can be found in (Schroeder 1998) for automobile evaluation and in (Kaminski 1996) for college selection.

Section 2.3: Further elaboration of strategic thinking in design appears in (Dym and Levitt 1991a). More detailed descriptions of the formal design methods can (also) be found in (Cross 1989), (Dym 1994a), (French 1985, 1992), (Pahl and Beitz 1984), and (VDI 1987). A strongly related discussion of concurrent engineering can be found in (Carlson-Skalak, Kemser, and Ter-Minassian 1997). More detailed descriptions of the means for acquiring and processing knowledge are (also) given in (Bovee, Houston, and Thill 1995), (Ulrich and Eppinger 1995), and (Jones 1992).

Section 2.4: The fundamentals of organizing teams are explained in (Tuckman 1965) and cited in (Bartol 1992). The discussion of Mary Parker Follett's explication of constructive conflict is adapted from (Graham 1996). The basic stages of group formation are discussed in most contemporary management texts.

Section 2.5: The two Xela-Aid chicken coop designs that we use throughout as illustrative examples are derived from (Gutierrez et al., 1997) and (Connor et al., 1997).

2.7 EXERCISES

2.1 Describe in your own words the similarities and differences between the four models of the design process shown in Figures 2.1–2.4.

2.2 When would you be likely to use a descriptive model of the design process? When would you use a prescriptive model?

2.3 Map the management process shown in Figure 2.5 onto the design process shown in Figure 2.4.

2.4 Explain the differences between tasks, methods, and means.

2.5 You work for HMCI, a small engineering design company. You have been named team leader for a four-person design project that will be described in more detail in Exercises 3.2 and 3.5. You have not previously worked with any of these team members. Describe several strategies for moving the team quickly to the performing stage of group formation.

2.6 As Director of Engineering at HMCI, you notice that one of your team leaders and the team's client are unable to agree on a schedule. How might you advise the team leader to resolve this matter contsructively?

Chapter 3

Understanding the Client's Problem

What does this client want?

In the preceding three chapters we have defined engineering design, explored and described the process of design, and detailed some tools that we can use to monitor and control a design project. Now we turn to describing the tools we use in the preprocessing phase of design, during which we are working toward developing an engineering definition of the problem. Indeed, the set of activities done during the preprocessing phase is also known as *problem definition*.

3.1 OBJECTIVES TREES: TRANSLATING AND CLARIFYING THE CLIENT'S WANTS

The starting point of most design projects is the identification by a *client* of a *need* to be met. The fulfillment of that need then becomes the goal of the chosen design team. The client's need, as depicted in our models of the design process pictured in Figures 2.1–4, is quite often presented as a verbal statement in which the client identifies a gadget that will appeal to certain markets (e.g., a container for a new beverage), a widget that will perform some specific functions (e.g., a chicken coop), or a problem to be fixed through a new design (e.g., a new transportation network and hub).

Sometimes clients' project statements are quite brief. For example, consider the beverage container problem described in Section 2.5. Whether the design team is working for Great American Food and Tobacco (GRAFT) or for Bringing Juice

Into Children (BJIC), it would likely get a memo from upper management that says: "Design a bottle for our new children's fruit juice product." The design team could easily respond to this directive by choosing an existing bottle, slapping on a clever label, and calling its work done. However, we might ask whether this new bottle is a *good* design, or, further, whether it's the *right* design. Answers to these questions depend on how we measure goodness and on how we assess rightness or correctness. We will discuss metrics against which we can measure designs in Chapter 5.

Another simple project statement might take the form of "The Claremont Colleges need to reconfigure the intersection of Foothill Avenue and Dartmouth Avenue so students can cross the road." While communicating someone's idea of what the problem is, statements like this one have limitations because they often contain errors, show biases, or imply solutions. *Errors* may include incorrect information, faulty or incomplete data, or simple mistakes regarding the nature of the problem. Thus, the problem statement just given should refer to Foothill Boulevard, not Foothill Avenue. *Biases* are presumptions about the situation that may also prove inaccurate because the client or the users may not fully grasp the entire situation. In the traffic example, for instance, the real problem may not be related to the design of the intersection but to the timing of the signal lights or to the tendency of students to jaywalk. *Implied solutions*, that is, a client's best guesses at solutions, frequently appear in problem statements. While implied solutions offer some useful insight into what the client is thinking, they may wind up restricting the design space in which the engineer searches for a solution. Also, sometimes the implied solution fails to actually solve the problem at hand. For example, it is not obvious that reconfiguring the intersection will solve the student traffic problem. If students jaywalk, reconfiguring the intersection will do little or nothing to mitigate this. If the problem is that students are crossing a dangerous street, we may want to relocate the destination to which they are headed. The point is that we must carefully examine project statements in order to identify and deal with errors, biases, and implied solutions. Only then do we get to the real problem.

For now, we want to focus on developing a clearer understanding of what the client wants because this will help us see the lines along which measures for a design might emerge. That is, we want to clarify what the client wants, account for what potential users need, and understand the technological, marketing, and other contexts within which our gadget or widget will function. In so doing, we will be *defining our design problem* much more clearly and much more realistically. (We will see that this is where we start to think about what will emerge as the *product specifications*, the formal statements of the properties and functionalities that our design must have.)

3.1.1   Object Attributes and Objectives Trees

Imagine that we are members of a design team that is consulting for a company that makes both low- and high-precision tools (with corresponding prices). The company's management, seeking to penetrate a new market, has given the team a charter more specific than designing a "safe ladder," to wit, "Design a new ladder for electricians or other maintenance and construction professionals working on conventional job sites." This is a fairly "routine" design task, but to really understand the goals of this design, we still need to talk with management, some potential users, some of the company's marketing people, and some experts. We also need to conduct our own brainstorming sessions. We will get a better understanding of what our design project is really about by asking questions such as:

- What features or attributes would you like the ladder to have?
- What do you want this ladder to do?
- Are there already ladders on the market that have similar features?

And while asking these three questions, we will also want to ask:

- What does that mean?
- How are you going to do that?
- Why do you want that?

As a result of our discussions and brainstorming, we might generate the list of characteristics and attributes of a safe ladder design shown in List 3.1

List 3.1.   SAFE LADDER Attributes List

Ladder should be useful
Used to string conduit and wire in ceilings
Used to maintain and repair outlets in high places
Used to replace lightbulbs and fixtures
Used outdoors on level ground
Used suspended from something in some cases
Used indoors on floors or other smooth surfaces
Could be a stepladder or short extension ladder
A folding ladder might work
A rope ladder would work, but not all the time
Should be reasonably stiff and comfortable for users
Step deflections should be less than 0.05 in.
Should allow a person of medium height to reach and work at levels
    up to 11 ft
Must be safe
Must meet OSHA requirements
Must not conduct electricity
Could be made of wood or fiberglass, but not aluminum
Should be relatively inexpensive
Must be portable between job sites
Should be light
Must be durable
Needn't be attractive or stylish

We note in examining this list that not all of the statements are of the same kind. Some of them can be considered as binary issues, answered either by a "Yes" or a "No," while others allow for a range of answers. For example, the statement "Must not conduct electricity" doesn't really leave any options: Either the ladder is a conductor or it is not a conductor, and in this case it must be an insulator. On the other hand, statements such as "Should be relatively inexpensive" allow for a range of prices at which the ladder could be built and sold. A ladder that can be built for $15 is more desirable than one that can be built for $20, assuming all the other characteristics are the same.

There are also other differences in these statements. The material of which the ladder is to be made (wood or fiberglass, but not aluminum) is a design choice that should probably be deferred until later in the design process, unless there is some specific information from the client that forces an early choice. And the idea that the ladder must resist certain forces in certain ways (e.g., a limit on the deflection of a step) is a reflection of the way that engineers begin to translate features of a design into specifications, a subject we return to in Chapter 5.

The first step we take toward understanding the goals of a design project is building an objectives tree *for the object being designed.* An *objectives tree* is a graphical depiction of the objectives or *goals for the artifact* (as opposed to goals for a design project or process). The top-level goal in an objectives tree, which we represent as a node at the peak of the tree, is decomposed or broken down into subgoals that are at differing levels of importance or that include progressively more detail, so the tree reflects a *hierarchical structure* as it expands downward. An objectives tree also shows that related subgoals or similar ideas can be *clustered* together, which gives the tree some organizational strength and utility. In Figure 3.1 we show an objectives tree for a safe ladder design project such as the one we discussed in Chapter 2. The most interesting question about this tree may be, How did we construct that objectives tree?

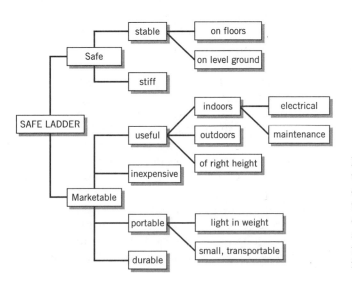

Figure 3.1 The objectives tree for the design of a safe ladder. It shows the first fruits of problem definition. Note the hierarchical structure and the clustering of similar ideas.

3.1.2   Goals and Objectives, Constraints, Functions, and Implementations

The reason that the statements in the foregoing list for the safe ladder seem different in kind is that they reflect different intellectual objects that must be considered and evaluated by a designer. There are clearly some *objectives* we want to achieve (e.g., the ladder should be relatively inexpensive), some *constraints* (e.g., a step should deflect no more than 0.05 in.), some *functions* (e.g., to reach auditorium light fixtures), and there are some *means* or *implementations* (e.g., the ladder could be made of wood or fiberglass). This suggests in turn that perhaps we need to categorize all of these elements to decide which ones go into an *objectives* tree, and in what order. We have already defined some of these terms in Section 1.2, but we will review them here (with the intent of being able to recognize them in the context of an attribute list having different kinds of statements that would emerge very early in a design project, such as the safe ladder list just given).

Goals and *objectives* are ends that we strive to achieve by exerting some effort, by making something happen. (We generally view design objectives and design goals as meaning much the same thing, with the possible exception that reference to *the* design goal as the top-level objective.) Objectives are usually expressions of the desired attributes and behavior that the client or potential users would find attractive. They are normally expressed as "being" statements that say what the design will *be*, as opposed to what the design must *do*. For example, saying that the ladder should be stiff is a "being" term. "Being" terms identify attributes that make the object look good in the eyes of the client, expressed in the natural languages of the client and of potential users.

Another way of identifying objectives is to note that they are often written as statements that "more (or less) of [the goal]" is better than "less (or more) of [the goal]." For example, lighter is better than heavier if our goal is portability. As such, objectives lend themselves to being measured somehow. In this way, objectives help us to choose among alternative design configurations.

Constraints are restrictions or limitations on a behavior or a value or some other aspect of a designed object's performance. Constraints are typically stated as clearly defined limits whose satisfaction can be framed into a binary choice, for example, the ladder material is a conductor or it is not, or the step deflection is less than 0.05 in. or it is not. Constraints are important to the designer because they limit the size of a design space by forcing the exclusion of unacceptable alternatives. For example, any ladder design that fails to meet OSHA standards will be rejected.

Objectives and constraints sometimes seem to be interchangeable, but they are not. On the other hand, objectives and constraints are closely related. Constraints limit the size of the design space, while objectives permit the exploration of the remainder of the design space. That is, constraints can be formulated so as to allow us to reject alternatives that are unacceptable, while objectives allow us to select among design alternatives that are at least acceptable, or, in other words, they *satisfice*. Designs that satisfice (or alternative selections in situations that require a choice to be made) may not be the best or optimum, but they do, at least, minimally satisfy all constraints. For example, we could barely satisfy OSHA standards or we could satisfy them "in spades" by making a "super safe" ladder in order to obtain a marketing advantage. Or, on the price side, a goal that a ladder should be "low cost" could be cast as a statement that the cost to build the ladder cannot exceed $25. In that case we have a fixed limit, a constraint. If we have *both* a low-cost objective and a $25 constraint, we may be able to exclude some initial designs based on the constraint alone, leaving us free to choose among the remaining designs based on cost.

It is important to recall that both objectives and constraints refer to the object being designed, not to the design process. For example, while rejecting a ladder design for failing to meet OSHA standards is applying a constraint, designing a ladder to meet OSHA standards should not be entered as a goal in the objectives tree. This is because identifying the OSHA constraint with a wish to meet OSHA standards is mistaking objectives for the artifact with objectives for the design process. As we have already noted, an objectives tree is *about the designed object* because it incorporates what both the client wants and the user(s) need from the design. We will have some more things to say about constraints and objectives in Section 3.2

Functions are the things a design is supposed to *do*, the actions that it must perform. In our initial attributes list, functions are usually expressed as "doing" terms that often reflect the language of the engineer. We will discuss functions in greater detail in Chapter 5.

Lastly, *implementations* or *means* are ways of executing those functions that the design must perform. These are the items on the attributes list that provide specific suggestions about what a final design will look like or be made of (e.g., the ladder will be made of wood or of fiberglass), so they often appear as "being" terms. However, it is usually pretty obvious which "being" terms are goals to be achieved and which "being" terms are very specific properties. It is premature for us to consider means in great detail here, since any means or implementations that we might select would be governed by the things that a specific designed object must do. That is, implementations and means are very much *solution-dependent* in that they are often design choices made to implement the functions that are to be performed by the already-chosen design.

Now that we have defined and identified the different kinds of attributes and characteristics on our initial list, we can remove the constraints, functions, and implementations, leaving only goals on the list. Since we are building an objectives tree, we are thus also pruning the tree of constraints, functions, and means. Thus, for the ladder, our pruned list of objectives is given in List 3.2.

While List 3.2 is useful as a list of goals to be achieved, there is much more that we can do with it. In particular, if our list was much longer, we might find it difficult to use the list without organizing it in some way. Consider the several uses that we have identified for the ladder. While this is not an exhaustive list of ways to use a ladder, we may want to group or *cluster* these uses together in some coherent way. And one way to start grouping entries on the list is to ask ourselves why we care about them. For example, why do we want our ladder to be used outdoors? The answer is probably because that's part of what makes a ladder useful, which is another entry on our list. Similarly, we could ask why we care whether the ladder is useful. In this case, the answer is not on the list—we want it to be useful so that people will buy it. Put another way, usefulness makes a ladder marketable. This suggests that we need an item on our list about marketing, for example, "The ladder should be marketable." This turns out to be a very helpful objective, since it tells us why we want the ladder to be cheap, portable, etc. If we want to cluster our questions this way, we will find a new list that we can put in the form of an *indented outline*, with *hierarchies* of major headings and various degrees of subheadings, as shown in List 3.3.

List 3.2.   SAFE LADDER Pruned Objectives List

Ladder should be useful

Used to string conduit and wire in ceilings

Used to maintain and repair outlets in high places

Used to replace lightbulbs and fixtures

Used outdoors on level ground

Used suspended from something in some cases

Used indoors on floors or other smooth surfaces

Should be reasonably stiff and comfortable for users

Should allow a person of medium height to reach and work at
    levels up to 11 ft.

Must be safe

Should be relatively inexpensive

Must be portable between job sites

Should be light

Must be durable

As we can see, this revised, indented outline allows us to explore each of the top-level objectives further, in terms of the subobjectives that tell us how to realize it. At the highest level, our objectives turn us back to the original design statement we were given, namely to design a safe ladder that can be marketed to a particular group.

Now, we have certainly not exhausted all the questions we could ask about the ladder, but we can identify in this outline some of the answers to the three questions mentioned just above. For example, "What do you mean by safe?" is answered by two subgoals in the cluster of safety issues, that is, that the designed ladder should be both stable and relatively stiff. We have answered "How are you going to do that?" by identifying several subgoals or ways in which the ladder could be useful within the "The ladder should be useful" cluster and by specifying two further "sub-subgoals" about how the ladder would be useful indoors. And we have answered the question "Why do you want that?" by indicating that the ladder ought to be cheap and portable in order to reach its intended market of electricians and construction and maintenance specialists.

We can represent the indented outline we have just started in graphical form simply by laying out a hierarchy of boxes, just as shown in Figure 3.1, with each layer or row of boxes corresponding to a level of indentation (which is indicated by the number of digits to the right of the first decimal point) in the outline. Thus, our indented outline becomes an objectives tree. In fact, this process has a lot in common with one of the fundamental skills of writing, being able to construct an outline. A topical outline provides an indented list of topics to be covered, together with the details of the subtopics corresponding to each topic. Since each topic represents a goal for the material to be covered, the identification of an objectives tree with a topical (or an indented) outline seems pretty obvious.

List 3.3.   SAFE LADDER Indented Objectives List

0. *A safe ladder for electricians*
   1.   The ladder should be safe
      1.1   The ladder should be stable
         1.1.1   Stable on floors and smooth surfaces
         1.1.2   Stable on relatively level ground
      1.2   The ladder should be reasonably stiff
   2.   The ladder should be marketable
      2.1   The ladder should be useful
         2.2.1   The ladder should be useful indoors
            2.2.1.1   Useful to do electrical work
            2.2.1.2   Useful to do maintenance work
         2.2.2   The ladder should be useful outdoors
         2.2.3   The ladder should be of the right height
      2.2   The ladder should be relatively inexpensive
      2.3   The ladder should be portable
         2.3.1   The ladder should be light in weight
         2.3.2   The ladder should be small when ready for transport
      2.4   The ladder should be durable

Still further, the graphical format of the tree is quite useful for discussions with clients and other participants in the design process. It is also useful for determining what things we need to measure, since we will use these objectives to decide among alternatives. The graphical format or tree is also useful since it corresponds to the mechanics of the process that many designers follow. Often, the most useful way of "getting your mind around" a large list of objectives is to put them all on Post-It™ notes, and then move them around until the design team is satisfied with the tree. We will discuss some of the mechanics of tree building and problem definition in Section 3.5.

One final point about this simple example. Note that as we work *down* the tree, or move further in on the levels of indentation, we are doing more than just getting into more detail. We are also answering a generic *how* question for many aspects of the design, that is, the question of "*How* are you going to do that?" Thus, as we get further and further down the tree, we are beginning to be able to identify the *functions* that our designed object must perform.

Conversely, as we move *up* the tree, or further out toward fewer indentations, we are answering a generic *why* question about a specific (needed) function, that is, the question of "*Why* do you want that?" This enables us to track why we want some feature or other fine point in our design, which may be very important if we have to trade off features, one against the other, because the values of these features may be directly attributable to the importance of the goals they are intended to serve. We will say more about this in Section 3.3.

3.1.3   How Deep is an Objectives Tree?
What About Pruned Entries?

Where do we end our list or tree of objectives? One simple answer is to stop when we run out of objectives or goals and begin to see functions appearing. That is, within any given cluster, we could continue to parse or decompose our subgoals until we are unable to express succeeding levels as further subgoals. The argument for this approach is that it points the objectives tree toward a *solution-independent* statement of the design problem. That is, we know what characteristics the design has to exhibit, without having to make any judgment about how it might get to be that way. In other words, we determine the attributes of the designed object without specifying the things the design must do that, in turn, may be strongly suggestive of a particular solution.

Another way of limiting the depth of an objectives tree is to look out for verbs or "doing" words because they normally suggest functions. Still further, when clearly identifiable implementations or means are listed, we should recognize that as a definite signal that we need to backtrack one or more levels from those means to the functions with which they are associated. Then both the "offending" functions and their implementations can be pruned together.

A second tree-building issue is deciding what to do with the things that we have removed from the list. In the case of the functions and implementation, we simply put them aside (recording them in case they are good ideas), and pick them up again later in the process. In the case of constraints, however, it is reasonable to reenter them into an appropriate place in the objectives tree, while being careful to distinguish them from the objectives. For example, in an outline form of the objectives tree, we might use italics or a different font to denote constraints (see List 3.4 in the next section). In a graphical form, we may wish to highlight constraints using differently shaped boxes. In either case, it is important to recognize that constraints are related to but different from objectives, and they are used in different ways.

3.1.4   The Objectives Tree for the Beverage Container Design

In the beverage container design problem, our design team is working for one of the two competing food product manufacturers, in this instance BJIC. (We note parenthetically that there is an interesting ethical problem that we will address in Chapter 9, that is, Could our design team, or our firm, take on the same or similar design tasks for both, or for two competing clients?) However, for now, let us suppose that we're dealing with a single client and that our client's project statement is as stated in Section 2.5, that is, "Design a bottle for our new juice product."

In order to clarify what was wanted from this design, our design team questioned many people in BJIC, including the marketing staff, and we talked to some of their potential customers or users. As a result, we found that there were several motivations driving the desire for a new "juice bottle," including: plastic bottles and containers all look alike; the client, as a national producer, has to deliver the product to diverse climates and environments; safety is a big issue for parents whose children might drink the juice; many customers, but especially parents, are concerned about environmental issues; the market is very competitive; parents (and teachers) want children to be able to get their own drinks; and, finally, children always spill drinks.

These motivations emerged during the questioning process, and their effects are displayed in the augmented attributes list for the container given as List 3.4. Some of the entries in this list are shown in italics because they are constraints, means, or implementations. Thus, these entries can be removed from a final list of the attributes that are objectives.

List 3.4.   BEVERAGE CONTAINER Augmented Attributes List

Safe	➡ DIRECTLY IMPORTANT
Perceived as Safe	➡ Appeals to Parents
Inexpensive to Produce	➡ Permits Marketing Flexibility
Permits Marketing Flexibility	➡ Promotes Sales
Chemically Inert	➡ *Constraint* on Safe
Distinctive Appearance	➡ Generates Brand Identity
Environmentally Benign	➡ Safe
Environmentally Benign	➡ Appeals to Parents
Preserves Taste	➡ Promotes Sales
Easy for Kids to Use	➡ Appeals to Parents
Resists Range of Temperatures	➡ Durable for Shipment
Resists Forces and Shocks	➡ Durable for Shipment
Easy to Distribute	➡ Promotes Sales
Durable for Shipment	➡ Easy to Distribute
Easy to Open	➡ Easy for Kids to Use
Hard to Spill	➡ Easy for Kids to Use
Appeals to Parents	➡ Promotes Sales
Chemically Inert	➡ *Constraint* on Preserves Taste
No Sharp Edges	➡ *Constraint* on Safe
Generates Brand Identity	➡ Promotes Sales
Promote Sales	➡ DIRECTLY IMPORTANT

The augmented list (List 3.4) also shows how, after additional brainstorming and questioning, some of the listed goals are either expanded into subobjectives (or subgoals) and others are connected to existing goals at higher levels. In one case a brand new top-level goal, Promote Sales, is identified. The objectives tree corresponding to (and expanded from) this augmented attribute list is shown in Figure 3.2, and a tree combining objectives and constraints is shown in Figure 3.3. The detailed subgoals that emerge in these trees clearly track well with the concerns and motivations identified in the clarification process.

As a result of the thought and effort that went into List 3.4 and the objectives trees of Figures 3.2 and 3.3, the design team rewrote and revised the problem statement for this design project to read: "Design a safe method of packaging and distributing our new children's juice product that preserves the taste and establishes brand identity to promote sales to middle income parents." Thus, as we noted in Chapter 2, one of the outputs of the design preprocessing (or problem definition) phase is a revised statement that reflects what has been learned about the goals for a design project. That is, the emergence of a clearer understanding of the client's design problem results in an objectives tree that points toward the expression of the features and behaviors wanted from the designed object, and it often results in a simultaneous revision or restatement of the client's original problem statement.

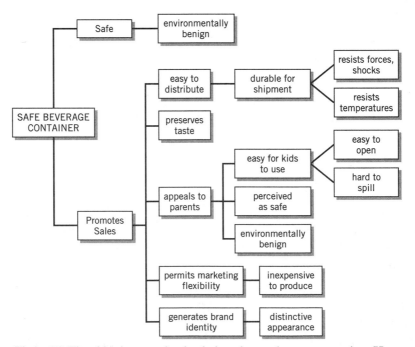

Figure 3.2 The objectives tree for the design of a new beverage container. Here the work on problem definition has lead to a hierarchical structuring of the needs identified by the beverage company and by the potential consumers—or at least the consumers' parents!—of the new children's juice drink.

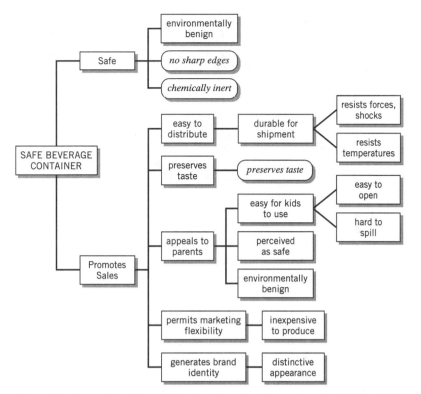

Figure 3.3 A combined tree (objectives in rectangles and constraints in ovals) for the design of a new beverage container. Here the goals for the new product are shown together with the constraints that apply to the object being designed.

3.2   CONSTRAINTS: IDENTIFYING LIMITS, OR WHAT CAN'T THE CLIENT HAVE?

There are limits to everything. That is why constraints are extremely important in engineering design, as we noted in Section 3.1.2 when we articulated some differences between constraints and objectives. Sometimes, as with OSHA standards for the ladder design, constraints that apply to the designed object are expressed as objectives for the design process itself, in which cases we must be careful not to confuse those objectives with goals for the designed object.

As a practical matter, many designers use constraints as a sort of "checklist" for designs in order to prune the set of designs to a more manageable size. Such constraints, which can be included in properly marked trees that include both objectives and constraints, are usually expressed as "hard," discrete numbers or points. By way of contrast, objectives are normally expressed as "soft," verbal "being" statements that can be formulated in terms of continuous variables or numbers. To reiterate our earlier illustration of this point, a goal that a ladder should be cheap could be stated in terms of having a cost that does not exceed a fixed limit or constraint, say $25. On the other hand, we could have *both* an objective that the ladder be cheap *and* a constraint that puts a limit on the cost. In that case we are able to choose among a set of designs whose costs to build are different, as long as they are all below the the limit set by the constraint alone. This, again, is the strategy of "satisficing" wherein we select among design alternatives that are acceptable.

There is still another approach to dealing with objectives that can be cast in terms of "continuous variables." There are many design domains in which we can formulate mathematical relationships between many of the design variables. For example, we might know how the cost of the ladder depends on its weight, its height, the size of its projected market, and other variables. In such cases, we may try to *optimize* or get the best design, say the minimum cost ladder, using procedures much the same as we use to find the maxima or minima of multi-variable calculus problems. Similarly, *operations research* techniques allow similar calculations to be performed when the design variables are discrete in nature, for example, when the ladder costs depend on the number of steps or connections, or if a ladder is restricted to be made in fixed-interval lengths, say that the ladder must be 5, 6, or 7 ft long. Optimization techniques are clearly beyond the scope of our discussions, but the underlying idea that design variables, and design objectives, interact and vary with one another, is also a theme that we will further elaborate in the following section when we discuss ways of assessing the comparative values of design goals.

3.3 ORDERING THE CLIENT'S WANTS: DEALING WITH SUBJECTIVE VALUES

We have been rather insistent in this chapter that we be careful to properly identify and list all of the client's objectives, taking care also to ensure that we don't mix up constraints, functions, or means with the goals set for the object being designed. But what if the identified objectives have differing values, whether for the client or for the users? After all, so far we have just assumed that all of the top-level objectives are equal in the eyes of all concerned, if only because we have made no effort to distinguish between them because of differences in perceived value. It is almost certain that some goals will be worth more than others, so we ought to be able to recognize that and deal with it. So, how are we going to do that?

3.3.1 Pairwise Comparison Charts: One Way to Rank Order Things

Suppose we have a set of goals for a project whose value we want to *rank*, that is, we want to identify their value or importance relative to one another, and to order them accordingly. Sometimes we get really lucky and our client expresses strong and clear preferences, or perhaps the potential users do, so that the design team doesn't have to do a serious ranking. On the other hand, oftentimes we do have to do some ranking or we have to place some values ourselves. Thus, we propose here a fairly simple technique that can be used to rank goals that are at the same level in the heirarchy of objectives and are within the same grouping or cluster, that is, they have the same parent or antecedent goal. (We will address more complex rankings when we describe weighted objectives trees in Section 3.3.3.) It is very important that we make our comparisons of goals with these clustering and hierarchical restrictions firmly in mind in order to ensure that we're comparing apples with apples and oranges with oranges. For example, would we really want to compare the subgoals of having the ladder be useful for electrical work and having the ladder be durable? On the other hand, rank ordering the ladder's usefulness, cost, portability, and durability would be good design information.

Suppose we are designing a laptop computer for which four high-level goals have been established: It should be low in cost, portable, convenient to use, and durable. Let us further suppose that we can easily choose between any given pair of them, for example, we prefer cost over durability, portability over cost, portability over convenience, and so on. Where would that leave us in terms of ranking all four goals? We can determine one answer to that question by constructing a simple chart or matrix that allows us to (1) compare every goal with each of the remaining goals individually and (2) add cumulative or total scores for each one of the goals. For example, in our four-goal problem such a *pairwise comparison chart* would look like:

Goals	Cost	Portability	Convenience	Durability	Score
Cost	••••	0	0	1	1
Portability	1	••••	1	1	3
Convenience	1	0	••••	1	2
Durability	0	0	0	••••	0

The entries in each box of the chart are determined as follows. Along the row of any given goal, say Cost, we enter a zero in those columns for the goals Portability and Convenience that are preferred over Cost, and we enter 1 in the Durability column because cost is preferred over durability. We also enter zeroes in the diagonal boxes corresponding to weighting any goal against itself, and we enter ratings of 1/2 for goals that are equally valued. The scores for each goal are determined simply by adding across each row, and so we see that in this case, the four goals can be ranked (with their scores) in order of decreasing value or importance, portability (3rd), convenience (2nd), cost (1st), and last, durability. We might wonder whether a score of 0 means that we can or should drop durability as an objective—and it takes only a little bit of clear thinking to see that we *cannot* drop objectives that score zeroes—but we'll discuss that in further detail in the next section.

There are other ways to rank order objectives, for example, every individual involved in the design process could assign points to each goal either on a preset scale (i.e., give each goal a value between 1 and 10) or assign each goal a numerical score such that the total number of points for all goals is fixed, say at 10 or

100 points. Then all of the individual scores obtained from, say, some of the stakeholders and members of the design team could be added up and averaged. However, it is clear that these approaches make individual or personal rankings even more subjective since every participant is ranking each goal against all of the others simultaneously, without the benefit of articulating the one-on-one choices inherent in the pairwise comparison chart.

It is important to keep in mind that the pairwise comparison method should be applied in a "top down" fashion, so that the higher-level objectives are compared and ranked before those at lower, more detailed levels. It seems only a matter of common sense to be sure that more"global" objectives (i.e., those abstract objectives that are higher up in the objectives tree) are properly understood and ranked before we fine-tune the details. For example, when we look at the objectives for the safe ladder (see Figure 3.1), it is more important to decide how we rank safety against marketability than its use for electrical work or for maintenance. Similarly, for the beverage container (viz., Figure 3.2), it is again more meaningful to rank safety against sales promotion before worrying about whether the container is easier to open than it is harder to spill.

It is also important to keep in mind that the pairwise comparison method, while more systematic than the two personal weighting approaches just described, is also subjective. Therefore, when we construct and use such ranking tools, we should ask whose values are being measured and incorporated. Marketing values could easily be included in different rankings, as in the ladder design, for example, where the design team might need to know whether it's "better" for a ladder to be cheaper or heavier. On the other hand, there could be deeper issues involved that, in some cases, may touch upon the fundamental values of both clients and designers. For example, let us turn to the beverage container design, now with the thought of looking at how the design objectives might be ranked if designs were being developed for the two competing companies, GRAFT and BJIC. We show the pairwise comparison charts for the GRAFT- and BJIC-based design teams in Figures 3.4 (a) and (b), respectively. It is clear from these two charts and the scores in their right-hand columns that the folks at GRAFT were far more interested in a container that would generate a strong brand identity and be easy to distribute than in it being environmentally benign or having appeal for parents. At BJIC, on the other hand, the environment and the taste preservation ranked more highly, thus demonstrating that subjective values show up in pairwise comparison charts and, eventually, in the marketplace!

3.3.2   Putting Subjective Rankings on a Scale

In addition to *ranking* objectives, or placing them in some order, it is often also imperative to *scale* them so that we can manipulate rankings in order to attach relative attaching weights to goals, as we will do in the next section. We need to weigh objectives one against another so that we can answer questions such as, *How much more* important is portability than cost in our laptop computer? Or, in the case of our beverage container, *How much more* important is environmental friendliness than durability? A little more? A lot more? Ten times more?

Figure 3.4 Pairwise comparison charts for the design of the new beverage container. Here, goals for the product are weighted one against another by designers working for (a) GRAFT and (b) BJIC. The relative values attached to each goal vary considerably in each chart, thus reflecting the different values held by each company.

Goals	Environ. Benign	Easy to Distribute	Preserves Taste	Appeals to Parents	Market Flexibility	Brand ID	Score
Environ. Benign	••••	0	0	0	0	0	0
Easy to Distribute	1	••••	1	1	1	0	4
Preserves Taste	1	0	••••	0	0	0	1
Appeals to Parents	1	0	1	••••	0	0	2
Market Flexibility	1	0	1	1	••••	0	3
Brand ID	1	1	1	1	1	••••	5

(a) GRAFT's weighted objectives

Goals	Environ. Benign	Easy to Distribute	Preserves Taste	Appeals to Parents	Market Flexibility	Brand ID	Score
Environ. Benign	••••	1	1	1	1	1	5
Easy to Distribute	0	••••	0	0	1	0	1
Preserves Taste	0	1	••••	1	1	1	4
Appeals to Parents	0	1	0	••••	1	1	3
Market Flexibility	0	0	0	0	••••	0	0
Brand ID	0	1	0	0	1	••••	2

(b) BJIC's weighted objectives

We can easily think of cases where one of the objectives is substantially more important than any of the others, such as safety compared to attractiveness or cost in an air traffic control system, and other cases where the objectives are essentially very close to one another. This sort of scaling can be done in a number of ways, including normalizing the rankings, setting the scores more or less arbitrarily in concert with our own views and then negotiating with the client or users, or using market research values. (In the last case, it is often the case that marketing people will argue that consumers are willing to pay a certain amount more for some quality.)

We can *normalize* our pairwise comparison scores or rankings as a basis for scaling, but only after we account for "zero-value" scores. For example, recall the score of 0 for durability in the pairwise comparison analysis we did for the laptop computer. Clearly we can't simply drop that objective because that says that this particular objective has no value at all compared to the other goals, or, equivalently, that the other goals are each worth infinitely more than the one with the zero score. It might also mean that if we performed such comparisons iteratively on the remaining goals, we'd be left at the end either with a set of goals that are equally valued or with a single goal. Such situations are not only uninteresting, they're also unrealistic because the world simply doesn't work that way.

Consider once more the simple four-goal comparison analysis of the laptop computer. A thoughtless analysis of the simple comparison chart of Section 3.3.1 would suggest that the four objectives of cost, portability, convenience, and durability be given normalized values of, respectively, $3/6 = 0.5$, $2/6 = 0.33$, $1/6 = 0.16$, and $0/6 = 0$. But, of course, this suggests that durability has no value to potential buyers of laptop computers, which clearly belies common sense. To account for the ordering (i.e., cost > portability > convenience > durability) and to get a sense of scale that includes all of the goals, we assign subjective values to the goals that are consistent with their rank and suggestive of how we might compare their relative importance. Thus, on a scale of 1 to 10, we could assign values to the four goals as follows: cost, 10; portability, 6; convenience, 3; and durability, 1. These values, within a fixed scale, identify the objectives that we deem the most important and the least important, and they place the remaining goals within that scale. There are several ways of normalizing these values, the easiest of which is to simply render the highest value as unity, that is, as 1, and then divide other values by 10. The values for the objectives for the laptop computer are, then: cost, 1.0; portability, 0.6; convenience, 0.3; and durability, 0.1.

Another way of dealing with zero-values is to use the "add one" method to change the range of assigned values that emerge from the pairwise comparison chart. The rankings for the laptop computer that emerged in Section 3.3.1 on a scale measured from 0 to 3 would be listed now as on a scale of 1 to 4. Thus, the four objectives of cost, portability, convenience, and durability are given normalized values of, respectively, $4/10 = 0.40$, $3/10 = 0.30$, $2/10 = 0.20$, and $1/10 = 0.10$. If we then normalized these laptop rankings to assign unity to the top objective, we would find: cost, 1.0; portability, 0.75; convenience, 0.50; and durability, 0.25. While these numbers look (and are) different from the previously assigned rankings, they preserve the same order and, as subjective results, are as valid (and as useful) as the scaling we have already obtained. The larger the number of objectives (or attributes or other choices) being ranked, the more closely the "add one" rankings would resemble our previously obtained normalized values that retained the rating of 0.

Values can also be normalized with respect to different normalization factors. For example, one normalization factor could be the arithmetic total of the total points awarded to all the values, which in this case is 20. Thus normalized, the (original) values for the objectives for the laptop computer are: cost, 0.5; portability, 0.3; convenience, 0.15; and durability, 0.05. While these normalized values look different than those normalized against the maximum individual value, their pair-by-pair ratios are unchanged. Similarly, we could also have normalized by dividing by the square root of the sum of the squares of the individual values, or, here, $\sqrt{(10)^2 + (6)^2 + (3)^2 + (1)^1} \cong 12.1$. In this case, the values for the objectives for the laptop computer are, then: cost, 0.83; portability, 0.50; convenience, 0.25; and durability, 0.08. Once again, the range of values has a different look because we've used a different normalizing factor, but the pair-by-pair ratios remain, as we should expect, unchanged.

One last point about scaled and normalized values. These are numbers that in some sense approximate subjective views or judgments about relative value or importance. Therefore, we needn't—and likely shouldn't—try to make these numbers seem more important by giving them unwarranted precision. That is, there is almost never a reason to write these normalized values as numbers that have more than two significant figures.

3.3.3   Weighted Objectives: What Does This Client Really Want?

We noted in the last section that it is often useful to integrate the (scaled and normalized) scores of the pairwise comparison charts into our objectives trees. This helps us to construct *weighted objectives trees* that explicitly show relative scores for every goal and subgoal. This display enables us to assess the relative importance or value of goals and subgoals all the way down through the tree, which in turn enables us to assess the relative value of every single subgoal that we're trying to realize in a given design. The weighted objectives tree also sets the stage for evaluating design alternatives, as we will detail in Chapter 6.

There are two key points to weighting objectives for a weighted objectives tree. The first is that every goal in the tree is assigned two numbers that are shown in the left- and right-hand boxes below each goal: The left-hand number is the goal's value relative to all of its cluster of goals at the same level in the tree. The right-hand number is sometimes referred to as the *true* value of that goal relative to the top-level goal for the object being designed. The true value for each goal in the tree is calculated as the product of its value relative to the other goals at its level (its left-hand number) and the true value of the parent goal from which that goal is descendent (the right-hand number of its parent goal). A simple illustration of how this weighting calculation works can be seen in Figure 3.5, wherein we can think of each cell as a goal with corresponding relative and true values. Note that the true, right-hand values of all of the subgoals in the last row of goals in the tree below totals to 0.33 + 0.33 + 0.22 + 0.11 = 0.99 ≅ 1.

The second key point is that the left-hand relative numbers given for each objective in the tree are derived from normalized versions of the scores or weights in the corresponding pairwise comparison chart. We have shown a more extended version of a weighted objectives tree for the beverage container design in Figure 3.6. The

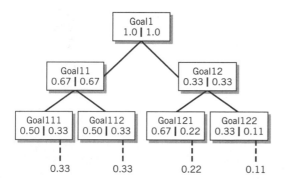

Figure 3.5 Constructing a weighted objectives tree. Two numbers are given for each of the goals: On the left is the value of that goal relative to all others at the same level within that cluster, while on the right is the true value for that goal relative to the top-level goal.

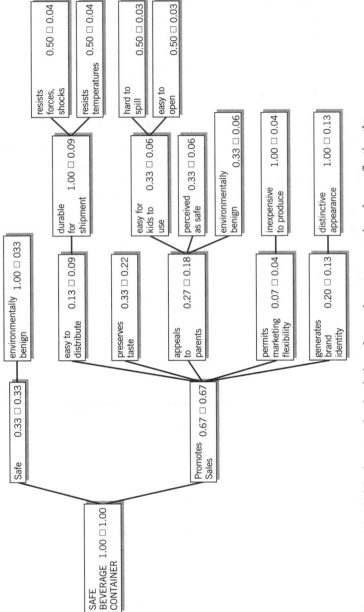

Figure 3.6 A weighted objectives tree for the design of a new beverage container, here reflecting the values of the BJIC company.

scores or weights used therein are obtained from the values expressed in the pairwise comparison chart of the BJIC company. Those values are clearly more environmentally and kid-friendly, but there are tradeoffs to be made even for BJIC's design.

As a practical matter, normalized values of the weighted objectives are used more often as a starting point for discussions with clients than as an end in themselves. In many cases, these numerical values cause clients or users to reflect upon their stated values and perhaps change them.

We also want to add a further note about the depth of weighted objectives trees. It is often unrealistic to calculate the weights more than a few levels down into the tree. Calculating weights deep down might produce numbers that are only distinguishable if many significant figures are kept. In such cases, we might be better off to rank order the lower levels without scaling them. Having said this, it is worth noting that there are cases, particularly where formal public processes are involved, in which we may have to weight all the levels, even if it seems like a silly exercise. In siting or locating highways, for example, complete and thorough application of design processes is a key element in obtaining regulatory approvals. However, as engineering designers, we should always remember that these processes can enlighten and guide, but they cannot replace sound and experienced judgment.

3.4   DESIGNING CHICKEN COOPS FOR A GUATEMALAN WOMEN'S COOPERATIVE

The first engineering course that engineering majors take at Harvey Mudd College is called E4: Introduction to Engineering Design. In this course, described in greater detail in the Preface, freshman engineering students are assigned the task of developing a conceptual design for a device or system. The projects are typically done for the benefit of a nonprofit or educational institution, and they provide the students with the insight that good engineering design may be (and is) done in nontraditional, noncorporate settings. The course also stresses the formal design methods we are presenting in this book. In order to illustrate student design within the E4 environment, we now describe the design of a chicken coop for use in the remote Guatemalan village of San Martin Chiquito. The sponsor of this project, Xela-Aid, is a humanitarian organization committed to working with and for people in Guatemala. Xela-Aid has a long history of working with the students in Harvey Mudd's E4 course. Among the E4 projects that have been done for Xela-Aid are designs for greenhouses, playgrounds, and improved methods for carrying burdens up a steep and treacherous mountain roadway.

In the design problem at hand, E4 teams were asked to design a chicken coop that would increase egg and chicken production, using materials that were readily available and maintainable by local workers. The end users were to be the women of a weaving cooperative who wanted to increase the protein in their children's diet in ways that are consistent with their traditional diet, while not appreciably distracting from their weaving. An extended version of the initial problem statement given in Section 2.5 appears as follows:

> The women of this village currently raise chickens in small fenced areas, but the few eggs and chickens produced must be sold at market to supplement the income of the family. Currently, chickens are kept in a small, fenced area (12 ft × 12 ft) with nothing but food on the ground and water bowls inside. Many of the eggs become cracked, and the coop floor becomes lower as the waste is swept out with a broom. Currently, the waste is placed in a shallow ditch. With newly purchased land, the women hope to be able to increase production, and have agreed that half of all the eggs produced will be used for their families, the other half, sold. The challenge will be to raise the most chickens/eggs possible in a small space of land, and in the most cost-efficient manner. Cold temperatures will require that the chickens are kept warm, but while electricity is available, relying on it could make the venture too costly. The women would like a low-upkeep system, as well, by which feed and water containers could be used that would require less filling, but in which the water will not putrefy, nor feed become moldy. A method for easy cleaning is also sought that will not erode the floor of the coop.

A reading of the above initial problem statement makes it clear that the design teams had many questions to answer before they could begin to specify the chicken coop's ultimate form. Probably the most pressing of these questions is, What exactly do the client and the end users want (and need), and what are the most important of the possible answers? To answer the first part of this question, the students had to undertake research into chicken husbandry (i.e., how to raise chickens), existing designs for chicken coops, and the cultural and climatic environment of Guatemala. In addition, teams had to determine what terms such as "cost-efficient" and "low-upkeep" meant to the client and to the end users. This was accomplished by a combination of library research, web searches, interviews with local (to Harvey Mudd) poultry producers and researchers, and repeated interviews with the client's liaison. The end result of this was the development of weighted and unweighted objectives trees and of refined client statements. We show in Figures 3.7 and 3.8 the objectives trees developed by two different teams. They are presented as elaborated by the student teams and, as such, have some errors or problems that are worth further exploration. For example, "noneroding floor" in Figure 3.7 is either a means or a constraint, but it is certainly not an objective. However, even with their shortcomings, there are a number of interesting points to be made about these two objectives trees. First, the two trees are clearly not identical. While this is not surprising, given that the trees reflect the work of two different teams, it serves to highlight the fact that many of the objectives, goals, and constraints that occur to designers are subject to analysis, interpretation, and revision. It is extremely important that engineering designers carefully review their findings with the client before proceeding too far in the design process. A second point to note is that one of the teams has

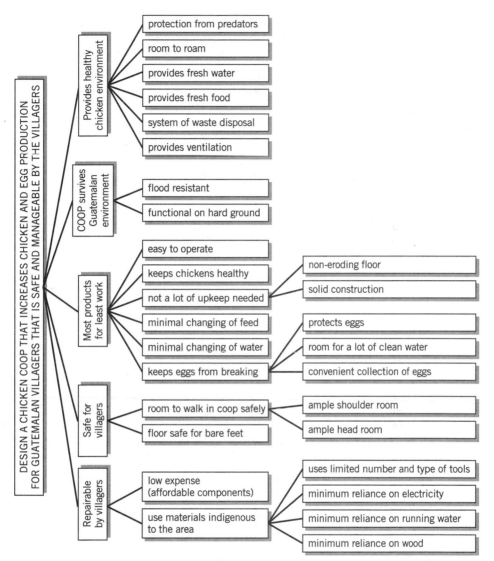

Figure 3.7 An objectives tree done by a freshman design team for the Xela-Aid chicken coop project. There are some entries in this tree that don't belong. Can you identify them?

chosen to incorporate the results of its research (e.g., "prevent egg cannibalism" in Figure 3.8), while the other has kept its objectives at a more general level, reflecting primarily what the client's liaison had indicated in personal interviews. The designer often informs and educates the client, providing the client with a better understanding of the problem as the process of clarifying objectives unfolds. This takes on a particular importance when we are considering functions and parametric specifications (cf. Chapter 5).

Figures 3.9 and 3.10 show the weighted objectives trees developed by the two design teams. Once again we encourage further review and consideration of

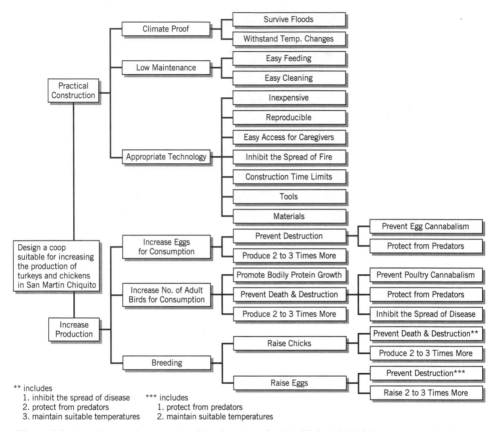

Figure 3.8 Another student team's objectives tree for the Xela-Aid chicken coop project. This tree also has some "objectives" that don't belong, including a performance specification (see Chapter 5). Note, too, that this team did not consider safety to be a top-level objective.

areas where they might be improved or changed in light of our discussions. Perhaps the most important issue in both cases is that of excessive detail. It is difficult to imagine many cases where weights that use three significant figures (e.g., 0.006 or 0.009) are useful in designing a prototype, particularly for a chicken coop! Here design teams will almost certainly want to revisit these numbers in some qualitative way (e.g., very important, important, nice but not important) or carefully consider how best to apply them in deciding among alternatives. We will discuss this topic in greater depth in Chapter 6.

After conducting their research and extensive interviews with their client, including several reviews of their unweighted and weighted objectives trees, the design teams revised their original problem statements and adopted new versions. One of the teams produced the following revised problem statement:

> The Xela-Aid organization would like our team to design and produce a chicken coop for use by several families in a small village in Guatemala. The climate of the village is tropical with temperatures rarely dropping

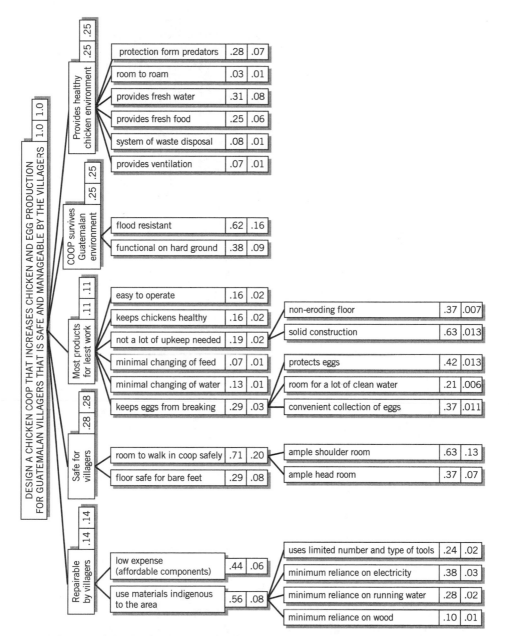

Figure 3.9 A weighted objectives tree for the Xela-Aid chicken coop project that corresponds to the objectives tree of Figure 3.7. Note that the numbers just "don't add up" to 1.00. Another concern is that the objectives are weighted to three significant figures, a degree of precision that is very hard to justify.

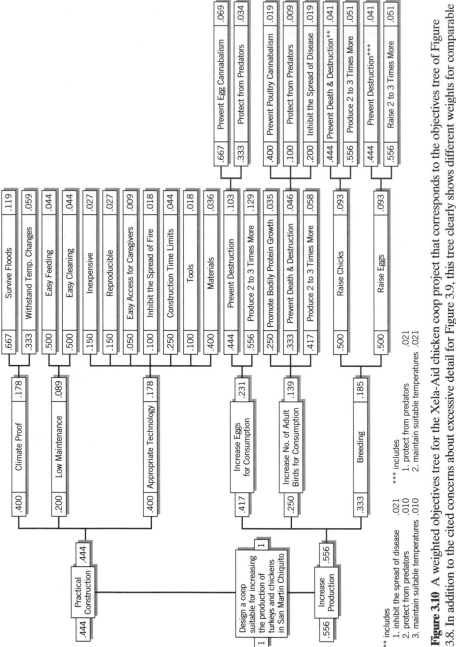

Figure 3.10 A weighted objectives tree for the Xela-Aid chicken coop project that corresponds to the objectives tree of Figure 3.8. In addition to the cited concerns about excessive detail for Figure 3.9, this tree clearly shows different weights for comparable objectives than it does for that counterpart. Thus, we wonder how a pairwise comparison of objectives inferred from this tree would differ from a rank ordering inferred from Figure 3.9.

** includes
1. inhibit the spread of disease   .021
2. protect from predators   .010
3. maintain suitable temperatures   .010

*** includes
1. protect from predators   .021
2. maintain suitable temperatures   .021

below 60°F. The village sits in a valley of mostly porous volcanic soil, is located at a height of approximately 7000 ft above sea level, and is surrounded by rain forest. During the rainy season, which lasts from March through September, the village receives an average of over an inch of rain each day. The chicken coop is to be located on a plot of land 8 ft wide by 20 ft long. This segment of land is a portion of a larger plot upon which will also be located a greenhouse, and a building where the women will produce assorted fabrics for sale. Materials that are readily available and should be considered for use when designing the coop are large segments of chicken wire, rebar, concrete, and locally made concrete blocks. Green wood is also available, but should be used sparingly as it is cut directly from the surrounding forest. Tools and any additional specific materials such as screws, nails, or any other inexpensive compact items will be bought by Xela-Aid but should also be considered scarce when designing the coop. Electricity is available in the village but it is expensive and unreliable and should not be incorporated in the design. The village does not have running water; the water must be carried by foot to the village from a source 30 minutes away. The coop will be constructed by the Guatemalans with the assistance of Xela-Aid members, so construction skills will not be an issue as long as the design is kept reasonably simple. The overall goal of the new coop is to allow the families to at least double their current chicken and egg production. This will allow the families to sell half of the chickens and eggs and consume the other half. Secondary goals include minimizing the labor required to maintain the coop and chickens, and utilizing the chicken droppings as fertilizer. To make the coop less labor intensive, steps should be taken to make it easy to clean and easy to supply with fresh food and water to minimize spoilage and putrefaction, respectively. The primary predator of the chickens in this part of the country is a small catlike animal able to dig under the current fence protecting the chickens; special precautions should be made to ensure this animal is unable to enter the enclosure.

3.5 SOME NUTS AND BOLTS OF DEFINING THE PROBLEM

In this section, we focus on some of the practical issues involved in trying to clarify and articulate what the client and users want from a design, including issues about how objectives (and constraints) trees evolve, who is asked, when they are asked, and how the information and knowledge is handled and tracked.

3.5.1 Questioning and Brainstorming

There are two kinds of activities that design teams can initiate, more or less contemporaneously, after they have been engaged by a client to undertake a design job. The first is asking questions of the client(s), and of others who might have varying degrees of interest in the design. These other stakeholders should include potential users and experts in the field. The experts can include people versed in any relevant technology or in other technical aspects, and marketing experts who are familiar with the market of users toward which a design is aimed. We have already identified (in Section 3.1) the kinds of questions that can be asked, but it is perhaps useful to keep in mind that what is intended is a collegial approach to soliciting information, not an adversarial approach that might put a respondent on the defensive.

It is also useful to be prepared when asking questions, in part to guide the focus of the questions, in part to ensure that respondents feel their time is not being wasted. This is very important if a program of structured interviewing of experts and/or users is envisioned because such interviews, or similarly detailed survey forms, will not produce useful or serious responses unless respondents feel that it's worth their time to answer a lot of questions, whether orally or in writing.

The second activity that design teams can initiate in the problem definition phase is brainstorming. As we noted in Chapter 2, brainstorming is a group effort in which new ideas are elicited, retained, and perhaps organized into some problem-relevant structure. When the team is brainstorming to identify desirable attributes or features, it is very important that its focus not shift to functions and means. While there will inevitably be some "nonresponsive" suggestions, the team should try to stick to the topic at hand, namely identifying goals and objectives, and perhaps constraints. One way a team can do this is for its leader to present statements of ideas with phrases such as, "A desirable characteristic of the object would be . . ." This will serve to remind the others of the team's focus. Of course, our preferred outcome is a list of attributes and characteristics that can be pruned as described in Sections 3.1.2 and 3.1.3. This list can then be refined into an indented list of objectives for the design, or, an objectives tree.

3.5.2 When and How Do We Build an Objectives Tree?

When do we build an objectives tree? Right away? As soon as the client has offered us the design job? Or, should we do some homework first and perhaps try to learn more about the design task we're undertaking?

There's no hard and fast answer to these questions, in part because building an indented list of objectives or an objectives tree is not a mathematical problem with an attendant list of initial conditions that must be met first. Also, building a tree is not a one-time, lets-get-it-done kind of activity. It's an iterative process, but one that surely should begin after the design team has at least some degree of understanding of the design domain. Thus, some of the questioning of clients, users, and experts should have begun, and some of the tree building can go on episodically while more information is being gathered.

One interesting feature of building an objectives tree is logistical in nature. How do we organize all that information, particularly if we're sitting around a conference room table and doing some intensive brainstorming? Surely we'd use blackboards or whiteboards, but how do we do all the clustering and the hierarchical organization while team members are throwing out ideas in a rapid, stream of consciousness fashion? One way is to use Post-It™ notes, which come in various sizes these days, and to simply write up individual notes for each entry on the list or in the tree. The notes can then be pasted on a board or display and later moved around as the team begins to organize the list of design attributes.

Two minor but important points. First, it is important that someone take notes during brainstorming sessions, in order to ensure that all suggestions and ideas are captured. It's always easier to prune out and throw away things than to recapture spontaneous ideas and inspirations. Second, after a rough outline of an objectives tree has emerged, it can be formalized and made to look pretty (and presentable) simply by using any standard, commercially available software package for constructing organization charts or similar graphical displays.

3.5.3 Problem Definition, Objectives Trees, and Revised Project Statements

Much of what we have been discussing in this chapter is about design preprocessing or problem definition. We had clearly indicated in Chapter 2 that the methods used in this phase of the design process include objectives trees, pairwise comparison charts, and weighted objectives trees. The output of this phase of the design process includes detailed (weighted) objectives, constraints, and revised problem statements. (Note there are still some more methods and outputs for the preprocessing phase, and we will take care of those in Chapter 5.) While we've discussed the methods and shown some of the outputs, we've not said very much about revised problem statements, so let's look into that now.

We have assumed all along that design projects would be initiated with a relatively brief statement drafted by the client to indicate what he seems to want. All of the methods and outputs we have described in this chapter are aimed at understanding and elucidating these wants, as well as accounting for the wants of other potential stakeholders. As we gather information from clients, users, and other stakeholders, our views of the design problem will shift as we expose implicit assumptions and perhaps a bias toward an implied solution. Thus, it is important that we recognize the impact of the new information we've developed and that we formalize it by drafting a revised problem statement that clearly reflects our clarified understanding of the design problem at hand. We saw such a revised problem statement as one of the emergent products of the beverage container design (viz., Section 3.1.4), and a comparison of the initial and revised problem statements for this project speaks very clearly to the notion of exposing more precisely what the client wants.

3.6 NOTES

Section 3.1: More examples of objectives trees can be found in (Cross 1994), (Dieter 1991), and (Suh 1990). The very important notion of *satisficing* is due to Simon (1981).
Section 3.2: Constraints are discussed in (Pahl and Beitz 1997).

Section 3.3: More examples of comparison charts are given in (Cross 1994) and (Ullman 1997). Additional examples of weighted objectives trees can be found in (Cross 1994) and (Deiter 1991).

Section 3.4: The results from the Xela-Aid chicken coop design project are taken from final reports ((Gutierrez et al., 1997) and (Connor et al., 1997)) submitted during the spring 1997 offering of Harvey Mudd College's freshman design course, E4: Introduction to Engineering Design. The course is described in greater detail in (Dym 1994b).

Section 3.5: Insights into the need to find and remove bias and implied solutions in assessing problem statements have been provided by Collier (1997).

3.7 EXERCISES

3.1 Explain the differences between biases, implied solutions, constraints, and objectives.

3.2 The HMCI design team established in Exercise 2.5 has been given the problem statement shown below. Identify any biases and implied solutions that appear in this statement.

> Design a portable electric guitar, convenient for air travelers, that sounds, looks, and feels as much as possible like a conventional electric guitar.

Revise the problem statement so as to eliminate these biases and implied solutions.

3.3 Develop an objectives tree for the portable electric guitar. (Someone will have to play the roles of client and users for this design project.)

3.4 Design a strategy for obtaining the weights for the objectives tree of Exercise 3.3.

3.5 The HMCI design team established in Exercise 2.5 has been given the problem statement shown below. Identify any biases and implied solutions that appear in this statement.

> Design a greenhouse for a women's cooperative in a village located in a Guatemalan rain forest. It would enable cultivation of medicinal preventive herbs and aid the villagers' diets. It would also be used to grow flowers that can be sold to supplement villagers' income. The greenhouse must withstand very heavy daily rains and protect the plants inside. The greenhouse must be made of indigenous materials because the villagers are poor.

Revise the problem statement so as to eliminate these biases and implied solutions.

3.6. Develop an objectives tree for the rain forest project. (Someone will have to play the roles of client and users for this design project.)

3.7 Design a strategy for obtaining the weights for the objectives tree of Exercise 3.6.

Chapter 4

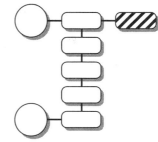

Managing the Design Process

You want it when?

In Chapter 2 we described the process of design, during which we identified and defined some formal methods of design. We also listed and defined some means of acquiring and processing the different kinds of knowledge and information needed to complete a design. In Chapter 3 we presented formal tools for articulating, understanding, and documenting what clients (and users) want from a design. It should be clear from what has been described so far that the design process is an activity that can consume significant amounts of time and resources. In this chapter we will explore some techniques that a design team can use to manage both its time and its other resources. That is, we will introduce ways of managing and controlling a design project. Our emphasis will be on describing the tools that can be used to successfully harness and organize all of the resources—including people, time, and money—needed by a design team as it strives to complete a design for a client under a variety of constraints, including both deadlines and budget limitations.

4.1 MANAGING DESIGN ACTIVITIES

In all of our discussions to this point, we have said very little about the context in which design activities occur. For example, there is a client to be satisfied, and there are regulations or laws that have to be obeyed. In point of fact, however, much of the environment in which design takes place is created by the designer, who must decide, among other things, which activities are to be performed, who is going to perform them, and the order in which they are to be completed. Thus, creating and controlling the design environment is part of a process that we know as managing design. In this chapter we will look at some of the tools that are available to the design team for planning, organizing, leading, and controlling design projects. Before we do so, however, it will be useful for us to consider briefly some of the aspects of design that make design projects hard to manage. In doing so, we will also see why careful application of appropriate tools can help the designer.

A project, whether intended to do design or aimed toward some other goal, can be characterized as "a one-time activity with a well-defined set of desired ends." Successful project management is usually judged in terms of scope, budget, and schedule. That is, the project must accomplish the goals (in our case a successful design), be completed within the resource limits available, and it must be done "on time." Consider for a moment our beverage container example. The design that is developed must meet the concerns of the beverage company, and these concerns could include developing an attractive, unbreakable container that is easily and cheaply made. If the project is being performed by a design firm, an agreed-upon budget must be met or the design firm may not be able to stay in business over the long run. The schedule might be dictated by marketing concerns, such as a new container being produced in time to sell the new juice in the coming school year. In this case, the scope of a successful design effort will balance all three sets of concerns, including the client's goal of introducing a new product, the budgetary constraints of the design firm, and the timing dictated by the client's marketing plan. On the other hand, if the project was being done by a student design team to fulfill a course requirement, budget concerns might then be measured in terms of student hours, given that students do have other commitments, including other courses, extracurricular activities, and so on. Further, the schedule would probably reflect the timing of the course, for example, that the design be done in a quarter or semester (or two). The scope of this version of the project would then address a combination of client, user, and faculty concerns. In either context, design firm or design course, it is the triad of *scope, spending, and scheduling*, the 3Ss, that provides the basis for the tools we use to manage projects.

Having noted the 3Ss of projects, we might ask whether design projects are different than other types of projects, such as construction projects, and if they are, how? It turns out that there are several important differences. The first is in the definition of a project's scope. For many projects, an experienced project manager knows exactly what constitutes a success. If it's a construction project to build a stadium, there are plans to be followed, including architectural renderings, detailed blueprints, and volumes of detailed parts and fabrication specifications. There are also accepted practices in the construction industry. As a result, the designation of a project manager implies that both the construction firm and the project manager understand the scope of this construction project. In a design project, on the other hand, the designer often cannot know what will constitute a success until the project is well underway. This is because she may not yet have held a full range of discussions with the client and with users, and even then may not be able to clarify all the objectives of the project and reconcile all the stakeholders' views. Similarly, while there is generally only one outcome that is expected from a construction project, many acceptable designs could be produced in a design project.

There are also differences in scheduling design projects, when compared with other projects. In a construction project, the project manager can plan to do certain activities, knowing how long each will take, and he can then determine a logical ordering for them. For example, digging a hole for a foundation might take two weeks, but it clearly must precede building forms (three weeks) and pouring concrete (two days) into thc holc. Assembling and organizing such planning and scheduling data allows the project manager to determine how much time the project will take. In many cases, the manager of a design project asks "How long do we have?" rather than summing up how long tasks will take. This approach to scheduling represents the design team's intent to use all the available time to generate and consider many viable design alternatives, while still trying to meet the constraint of finishing the project by some specified and agreed-upon time.

These differences might lead us to wonder if project management techniques are appropriate for use in engineering design projects. After all, if the scope seems ambiguous and the timing is up to the client and may even seem arbitrary, how can tools developed for well-understood projects be relevant? It turns out that the uncertainties and external impacts associated with design projects tend to make some of the management tools even more useful and necessary. As this chapter unfolds, we will see that project management tools can be useful for gaining agreement by the design team about what must be done, who will do it, and when things must be done—even if our expectations about the form of the final design are initially up in the air.

The team nature of design projects also lends support to the use of project management tools. We highlighted the stages of team formation that must be experienced in Chapter 2. There are some important concerns and issues that must be addressed when the team moves into the performing stage. These include the need to effectively communicate the activities, schedule, and progress of the team to all team members and to other stakeholders. We also need to allocate work fairly and appropriately. And we need to ensure that work has been done properly and in a sequence that allows other team members who depend on prior work to plan their actions. The management tools we introduce in this chapter are helpful in each of these circumstances.

4.2   AN OVERVIEW OF PROJECT MANAGEMENT TOOLS

Recall from Chapter 2 that we can model the process by which we eventually move from a client's problem to a detailed design for a solution. This process includes the use of a number of formal design methods and of means for gathering and organizing information in order to generate alternatives and evaluate their effectiveness. While the methods and means discussed can be assigned to the steps in the design process, design is not a simple "cookbook" process. Similarly, managing the design process also requires more than rote application of project management tools. In this section we briefly describe the tools that we can use to plan a design project, organize our design activities, agree on our responsibilities for the project, and monitor our progress.

In Chapter 1 we noted that management consisted of four functions: planning, organizing, leading, and controlling. *Planning* the project from a management perspective leads us immediately back to our 3S model of project requirements. We need to define the scope of the project, determine how much time we have to accomplish the scope (scheduling), and assess the level of resources (spending) we can apply to the project. *Organizing* the project consists primarily of determining who is responsible for each task area or activity of the project, and which other human resources can be called upon to work on the tasks. *Leading* a project means using tools to motivate a team by showing that the tasks can be understood, that the division of work is fair, and that the level of work can produce satisfactory progress toward the team's goals. Note, however, that project leadership cannot be provided by tools alone. Finally, *controlling* can only be done in the context created by the 3Ss and the plans that support those 3Ss. We can only track progress in a meaningful way against a stated set of goals, and we can change our plans or take corrective action if the team is confident that the plans they have developed are, in fact, going to be used.

The primary tool that is used to determine the scope of our activities is the *work breakdown structure* (WBS). The WBS is a hierarchical representation of all the tasks that must be accomplished to complete a design project. Project managers use WBSs to determine which tasks must be done. They will generally break the work down, hence the name, into pieces sufficiently small that the resources and time needed for each task can be estimated with confidence.

We use the *linear responsibility chart* (LRC) to determine which team member has the primary responsibility for the successful completion of each task in the WBS, as well as identify others who must participate in finishing that task. The LRC is basically a matrix that matches each of the tasks requiring management responsibility with the members of the team, the client, users, and other stakeholders. This is particularly important for team-based activities, both to clearly identify who is responsible for each task and to indicate any additional individuals (e.g., teammates, the client, or an external expert) that should be involved.

We can schedule activities in several ways, including using team calendars, Gantt charts, or activity networks. As its name implies, a *team calendar* shows all of the time that is available to the design team, with highlights that indicate deadlines and time frames within which work must be completed. A *Gantt chart* is a horizontal bar graph that maps various design activities against a time line. An *activity network* graphs the activities and events of the project, and it shows the logical ordering in which they must be performed. It is important to be aware that while a schedule can assist in planning and controlling the work of a design team, it can just as easily be little more than a flashy picture or marketing graphic.

The key tool we use to manage spending activities in a project is the budget. Essentially, a *budget* is a listing of all the items that will incur an economic cost, organized into some set of logically related categories (e.g., labor, materials, etc.). Note that there is an important distinction between the budget for doing the design, or design activities, and the budget needed to realize or build the artifact being designed. Our concern is primarily with the budget needed to do the design, and in Chapter 8 we will introduce issues of engineering economics and costing.

There are a number of control methods that are used to manage projects, but many of these are simply not well suited to design projects. For example, there is a technique called *earned value analysis* that relates costs and schedules to planned and completed work. While earned value analysis is useful for certain large-scale projects with effective reporting systems, it is overkill for the smaller, team-based projects that we discuss. A more useful and appropriate tool is the *percent-complete matrix* (PCM) that relates the extent of the work done to the total level of all work to be done. For the sake of simplicity, we have developed a version of this tool that is appropriate for smaller team design activities.

In the following sections we discuss and provide examples of each of these tools. Just as each of the methods and means given in previous chapters will not be applicable to every design project, not every team will need to use each of these management tools on every project. However, since they are important to good, team-based design work, and since we can never predict with certainty the kinds of design activities we will undertake in the future, these management tools are worth having in our individual arsenals.

4.3   WORK BREAKDOWN STRUCTURES: WHAT HAS TO BE DONE TO FINISH THE JOB

Most of us would be a little overwhelmed if we were asked to describe exactly how to start and drive a car, even though this is a fairly common task. In fact, we might say something like "first you get in the front seat." Of course, this presupposes that you know how to get into the car. If we were forced to describe this task to someone from a country with few cars and limited English-language skills, we might want to break down the task into a number of task groups, such as getting into the car, adjusting the seat and mirrors, starting the car, driving the car, and stopping the car. We might even want to review the entire plan before starting the car so that our driving student would know how to stop before starting out. This decomposition of tasks or concepts is the central idea of the work breakdown structure (WBS). When we are confronted with a very large or difficult task, one of the best ways to figure out a plan of attack is to break it down into smaller and more manageable subtasks. (Just as, and not so parenthetically, the way to learn to think about design or virtually anything else, is to decompose or break down the larger task into smaller and more easily defined subtasks.)

Perhaps a more believable example is that of a team that has been asked to design a spacecraft. That team will have to design across a number of specialties, including propulsion, communications, instrumentation, and structures. In this case the design team leader will work very hard to ensure that the team's propulsion experts are actually assigned to the tasks involving propulsion, that is, that the experts work on tasks relevant to their expertise. In order to do this properly, the team leader has to determine just what those tasks are. The WBS is a listing of all the tasks needed to complete the project, organized in a way that helps the project leader and the design team understand how all of the tasks fit into the overall design project.

We show a work breakdown structure for the beverage container design example in Figure 4.1. At the top level it is organized in terms of seven basic task areas:

- Understand customer requirements
- Analyze function requirements
- Generate alternatives
- Evaluate alternatives
- Select among alternatives
- Document the design process
- Manage the project
- Detailed design

We see also that each of these top-level tasks can be broken down in greater detail. Because of page size limitations, in this example we show great detail only for some of the tasks, such as understanding customer requirements. If we are actually part of a team carrying out this project, we would likely go into much greater depth in all the areas. Also, we should note that this method of organizing the work is not the only way to structure the WBS. We will see several alternative organizing frameworks later in this chapter.

Several observations about the WBS depicted in Figure 4.1 are in order. First, the basic principle for a WBS is that each item that is taken to a lower level is *always* broken down into two or more subtasks at that lower level. If the task is not broken down, then either the lower level is incomplete, or it is simply a synonym for the upper level. Second, a key rule for developing a WBS is that if we cannot determine how long an activity will take or who will do that activity, then we should probably break it down further. This suggests that experienced project managers will be more inclined to have shorter, less detailed WBSs than relatively inexperienced managers, since they are more likely to be able to aggregate subtasks into identifiable and measurable tasks just because of their greater experience.

Our third observation is that a WBS should be *complete* in the sense that any task or activity that consumes resources or takes time should be included either in the WBS explicitly or as a known component of another task. This is why the tasks of documentation and management are given in Figure 4.1. Activities such as writing reports, attending meetings, and presenting results are all necessary to the completion of the project, and failure to plan for them as work will certainly result in a poor outcome later. This requirement, estimating who and what is required and for how long, is a valuable discipline for any project, whether in a design course or in the "real world," both for developing the design and for ensuring that there is sufficient time to document and present results.

Our final observation on the WBS is that any part of its hierarchy of tasks should *add up*, that is, the time needed to complete an activity at a top level (e.g., Develop Objectives Tree) should be the sum of the listed tasks of the level below. This suggests that breaking work down to the next level below must be done in a thorough and complete way.

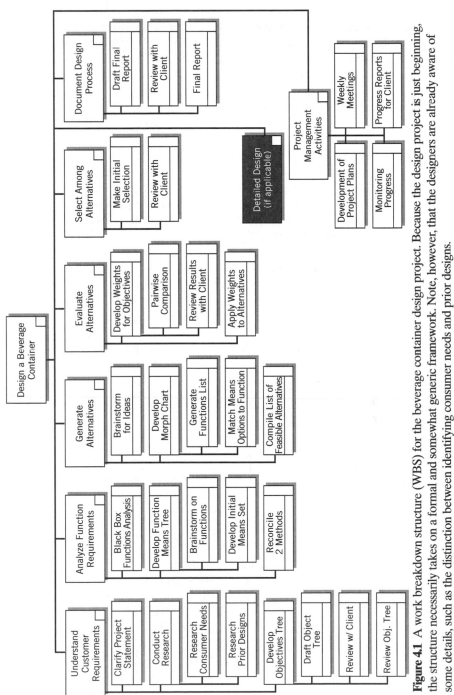

Figure 4.1 A work breakdown structure (WBS) for the beverage container design project. Because the design project is just beginning, the structure necessarily takes on a formal and somewhat generic framework. Note, however, that the designers are already aware of some details, such as the distinction between identifying consumer needs and prior designs.

The last two observations of the WBS thus provide us with two criteria for evaluating the utility of WBSs. *Completeness* refers to the idea that the WBS must account for all of the activities that consume resources or take time. *Adequacy* refers to the idea that tasks be broken down with sufficient quality, that is, to an adequate level of detail, such that the project team can determine how much time it will take to do them.

It is also important for us to note what the WBS is *not*. First, a WBS is *not* an organization chart for completing a project. This may be confusing to inexperienced project managers, since their previous experience with visually similar charts is often with "org charts." The WBS is a breakdown of tasks, not of roles or people in an organization. The second point is that a WBS is (also) *not* a flow chart showing the temporal or logical relationships among tasks. In many cases the listing of the tasks will be organized in such a way that a task (e.g., writing the final report) is shown in a different part of the hierarchy than other tasks which must precede it (e.g., all of the design, building, and testing that is being reported). Third and last, a WBS is *not* a listing of all the disciplines or skills that are required to complete the tasks. In many cases the tasks to be completed may require a number of different skills (e.g., electrical engineering and propulsion engineering). Tasks performed by professionals with these skills can be combined into the same part of the hierarchy if that listing of the tasks meets the above criteria of completeness and adequacy.

Figure 4.2 is another example of a WBS, in this case taken from an electrical hardware project. Note here that the project manager has chosen to organize the design

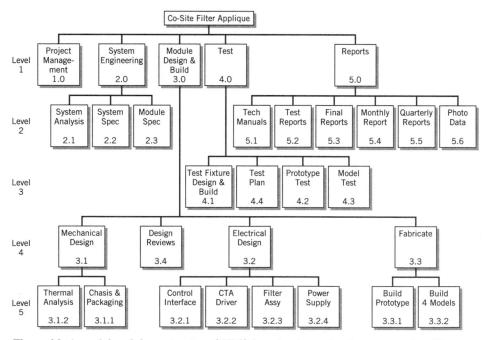

Figure 4.2 A work breakdown structure (WBS) for a hardware development project. Note that at the lowest level (level 5), the electrical design tasks have been broken down to the level of the components to be designed. At this stage, the designer of the Power Supply might follow another WBS similar to Figure 4.1. (After Kezsbom, Schilling, and Edward 1989.)

subtasks in terms of electrical and mechanical design. This may appear to violate our concern with disciplines noted above—until we realize that for this project, this is simply a convenient way to *organize the tasks*, which is always our key intent for a WBS.

Figure 4.3 is yet another example of a WBS, taken from a software package, Primavera Project Planner. In this case the example is for an automobile firm. Note that this WBS is not even in a graphical form, although it is still hierarchical. While we are inclined to use a graphical form for the sake of clarity, it is perfectly permissible to use a tabular form such as that shown in Figure 4.3. In fact, tables are a common means of collecting the information together once a WBS has been developed into its final form. Note also that this example has the work broken down by the various car components. This is also permissible, so long as the WBS satisfies our concerns about completeness and adequacy.

In the end, then, the WBS is a tool for a project team to use to make certain that they understand all the tasks that will be needed to complete their project. It is for this reason that it is so valuable for determining the scope of the project.

PRIMAVERA PROJECT PLANNER

Date 08JAN98   —WORK BREAKDOWN STRUCTURE—

ENGR–Active Projects for the Fiscal Year

Structure: xxx.xxx.xx.x

WBS Code   Title

94 All Projects
   94E   All Engineering Projects
      94E.101 Project E101
         94E.101.A General
         94E.101.A7
         94E.101.B Air Bag
         94E.101.C Mechanical Release System
         94E.101.D Electrical Systems
         94E.101.E Interior Dashboard
         94E.101.F Structural Door System
      94E.102 Retrofit Automobile Plant
         94E.102.A Enclosure
         94E.102.B Structural System
         94E.102.C Mechanical System
         94E.102.D Electrical System
         94E.102.E Estimating
         94E.102.F Specifications
         94E.102.G General
   94I   All Installation Projects
      94I.101   Tooling & Equipment Installation
         94I.101.A Structural Slab
         94I.101.B Piping
         94I.101.C Equipment
         94I.101.D Electricity
         94I.101.E Interior Finishes
         94I.101.F Ventilation & Plumbing
         94I.101.G General

Figure 4.3 A work breakdown structure (WBS) for the engineering projects of an automotive firm. This nongraphical WBS organizes the firm's activities according to systems for the autos and overall factory installation projects. The level of detail is not very high, and presumably the firm would have supporting WBSs for some if not all these projects.

4.4 LINEAR RESPONSIBILITY CHARTS: KEEPING TRACK OF WHO'S DOING WHAT

Once the tasks that are to be done have been identified through the WBS, a design team has to determine whether or not it has the human resources, the people, to accomplish those tasks,. The team also has to decide who will take responsibility for each task. This can be done by building a *linear responsibility chart* (LRC). The LRC lists the tasks to be managed and accounted for and matches them to any or all of the project participants. Figure 4.4 shows a simplified LRC corresponding to many of the tasks in the beverage container design project of Figure 4.1. In addition to all of the top-level tasks, the subtasks associated with several of the lower levels are also given. In practice, it is advisable to give all the top-level tasks, and those subtasks that may require management attention. As we implied earlier, this is a point where relatively inexperienced project managers would do well to provide more detail than is needed, rather than less.

As we can see in Figure 4.4, there is a row for each task within which the role, if any, of each project participant is given. These roles do not necessarily include assuming the primary responsibility. Indeed, most of the participants will play some sort of supporting role for many of the tasks, including reviewing, consulting, or working at the direction of whoever is responsible. A column is assigned to each of the participants, and this allows them to scan down the chart to determine their responsibilities to the project. For example, note that the client (or the client's designated liaison) will be called upon to give final approval to the objectives tree, the test protocol, the selected design, and the final report. The liaison must also be consulted during some of the predesign activities, and will be asked to review various intermediate work products. The client's research director is somewhat of a resource to the team, and may be consulted at different points of the project, but *must* be consulted regarding the test protocol. The team's boss, the director of design, has asserted the right to be kept informed at a number of points in the project, most notably in terms of design reviews. The project also has access to one or more outside experts, who are generally available for consultation, and who must review the design and certain other (unspecified) documents.

We also note from the LRC in Figure 4.4 that the team leader does not always have the primary responsibility for *every task* in the project. It is often the case in team projects that the team leader will not be responsible for tasks that are outside of her technical area of expertise, although she may want to specify a review or support role in order to remain informed. Sharing responsibility this way is sometimes quite difficult for teams and for team leaders. Sharing responsibility, which is strongly tied to the storming phase of group formation, requires practice. Therefore, the LRC can be used to make this part of team formation more explicit to the team, and to allow the team to reach consensus on who will be doing what in the project.

The LRC can also be used to let outside stakeholders in a project understand what they are expected to do. In the beverage container example, the client's research director clearly has an important role to play in the safe conduct of the

Figure 4.4 A linear responsibility chart (LRC) for the beverage container design project. Each participant in the project can read down his column and determine his responsibilities over the entire project. Alternatively, the Project Manager can read across a row and determine who is involved with each task.

Linear Responsibility Chart	Team Member #1	Team Member #2	Team Member #3	Team Member #4	Team Member #5	Director of Design	Client Liaison	Client Research Director	Outside Consultant
1.0 Understand Customer Requirements	1								
1.1 Clarify Problem Statement	1	2	2	2	2		3	4	4
1.2 Conduct Research	1	2			2		4	4	4
1.3 Develop Objectives Tree	1								
1.3.1 Draft Objectives Tree			2	2		5	3		4
1.3.2 Review w/ Client	1		2			5	5	3	4
1.3.3 Revise Objectives Tree	1		2	2			6		4
2.0 Analyze Function Requirements	2	2	1	2	2	5	4	3	3
3.0 Generate Alternatives				1					
4.0 Evaluate Alternatives	5	1	2	2	2				
4.1 Weight Objectives	1	2				5	6	3	3
4.2 Develop Test Protocol	5	1			2	5	4	5	3
4.3 Conduct Tests		1	2		2			5	
4.4 Report Test Results	5	2	2		1	5	5	5	
5.0 Select Preferred Design	1	2			2	5	6	4	4
6.0 Document Design Results		1							
6.1 Design Specifications	1			2		6			
6.2 Draft Final Report	5	1		2		5	5	3	4
6.3 Design Review w/ Client	1	2		2		5	3	4	3
6.4 Final Report	5			2	2	5	6	4	4
7.0 Project Management	1								
7.1 Weekly Meetings	1	2	2	2	2				
7.2 Develop Project Plan	1	2	2	2					
7.3 Track Progress	1					5			
7.4 Progress Reports	1						5		

Key:
1 = Primary responsibility
2 = Support/work
3 = Must be consulted
4 = May be consulted
5 = Review
6 = Final Approval

testing phase. It is very important that this person know early on what is expected and to be allowed to plan accordingly. Similarly, the outside experts may need to allocate time to ensure availability, and the director of design may need to make resources available to pay for the experts' time.

It should be clear from this that the LRC can be a very important document for translating the "what" of the WBS into the "who" of responsibility. At the same time, we might be tempted to include more into the LRC as a substitute for admitting that we or our team doesn't know something. For example, if every task has every team member assigned in a support or work role, it should raise serious doubts in our minds about whether we really do understand those roles. Similarly, if a team leader claims primary responsibility for all the tasks, her team will certainly be tempted to consider the LRC as little more than a power grab or a mirror of the team leader's insecurities. Thus, it is better to consider a matter open and leave its row blank than to fill it in blindly for the sake of a false conclusion. It is also important that a team understand that it may be necessary to revisit roles as the project unfolds, especially if the team is relatively inexperienced or the project is initially ambiguous.

4.5 SCHEDULES AND OTHER TIME MANAGEMENT TOOLS: KEEPING TRACK OF TIME

We use scheduling and similar time management tools to help us identify in advance those things that will really mess up our project if we don't get them done on time. There are three primary scheduling tools that are frequently used in project management: the calendar, the activity network, and the Gantt chart. A team calendar is probably the most familiar tool as it performs many of the same functions that our personal calendars or diaries do. It is simply a mapping of the deadlines or due dates of a project onto a conventional calendar.

The other two tools, the activity network and the Gantt chart, are both more powerful and, consequently, potentially more useful. Both are graphical representations of the logical relationships between tasks and the time frames in which they are to be accomplished. Indeed, most software programs for project management use the same information to generate both activity networks and Gantt charts. There are, however, important differences in practice that make it worthwhile to describe both so that the design team can decide which tool is most appropriate for its project.

4.5.1 Team Calendars: When Are Things Due?

As we have just said, a team calendar is simply a mapping of deadlines onto a conventional calendar such as might be found on a wall or desk. These deadlines will certainly include externally imposed ones, such as commitments to clients (or to professors in the case of academic projects), but should also include team-generated deadlines for the tasks developed in the WBS. In this sense, the team calendar is really an agreement by the team to assign the resources and time necessary to meet the deadlines shown on the calendar. Figure 4.5 shows a team

March 1999								Design Team		May 1999						
S	M	T	W	T	F	S				S	M	T	W	T	F	S
	1	2	3	4	5	6										1
7	8	9	10	11	12	13				2	3	4	5	6	7	8
14	15	16	17	18	19	20		**April 1999**		9	10	11	12	3	14	15
21	22	23	24	25	26	27				16	17	18	19	20	21	22
28	29	30	31							23	24	25	26	27	28	29
										30	31					

Sun	Mon	Tue	Wed	Thu	Fri	Sat
				1	**2** 5:00PM Prototype Built	**3**
4	**5**	**6** 7:00-8:15PM Team Meeting	**7**	**8**	**9** 11:00AM Proof of Concept Due	**10**
11	**12** 11:00AM Rough Outline Due	**13** 7:00-8:15PM Team Meeting	**14**	**15**	**16** 5:00PM Topic Stce Outline Due	**17**
18	**19** 11:00AM Prsntion Outline Due	**20** 7:00-8:15PM Team Meeting	**21** 11:00AM Slides Due	**22**	**23** 5:00PM Draft Final Report Due	**24**
25	**26** 10:00-11:00AM Present Results	**27** 7:00-8:15PM Team Meeting	**28**	**29**	**30** 5:00PM Final Report Due	

Figure 4.5 A team calendar for a student design project. Note that externally imposed deadlines, team commitments, and recurring meetings are all included on the calendar. It is usually better to make the team calendar "too complete" than it is to leave out potentially important milestones or deadlines.

calendar for a student design team that is seeking to complete its project by the end of April, in accordance with an externally imposed schedule. Note that the calendar includes a number of deadlines over which the team probably has no control, such as when the final report is due and when in-class presentation of results is to be done. It also includes routine or recurring activities, such as the Tuesday night team meetings. Finally, it includes some deadlines that the team has committed to realizing, such as completing a prototype by 5 P.M. on April 2.

In setting up a team calendar, several points should be kept in mind. First, the notion of the team calendar implies that the deadlines are all understood and agreed to by everyone on the team. As such, the calendar becomes a document that can—and should—be reviewed at each team meeting. Second, the team calendar should allow times that are at least consistent with the time estimates generated in the WBS. If a task was determined to take two weeks to complete, there is little point in allowing that task only one week on the team calendar. A final point to note is that the team calendar, while easily understood by members of the team, *cannot by itself capture the relationship between activities*. For example, in Figure 4.5 we see that building the prototype precedes proof of concept testing *only* because the team chose to put it that way. For many artifacts a proof of concept may actually precede building a final prototype. The team calendar cannot address this sort of problem, nor can it "remember" team decisions of this sort. For this reason, team calendars are really useful only for simple projects or in cases where they are supplemented with other project management tools (such as those we discuss in the following sections).

4.5.2 Activity Networks: Which Tasks Must Be Done First?

The activity network is based on the notion that each task and its output or result can be treated as a separate activity or event. Recall our earlier example of the construction project. We could consider the act of "digging a hole for the foundation" as an activity, and the "existence of the foundation hole" as an event. We could then consider the task "build forms" as an activity, and "forms erected" as an event. Finally, we could consider the task "pour concrete into forms" as another activity, and "foundation completed" as yet another event. Now, there are at least two ways that we could represent this set of activities and events graphically. For example, we could construct a network of nodes and connecting arcs, and then identify each node with an event and each arc as the activity that caused the event to happen. Alternatively, we could represent each activity with a node, and identify the events simply as the arcs that extend from the nodes. It obviously makes no logical difference which scheme we choose, *as long as we are consistent*. We will use the form of activity network called the *Activity-on-Node* (AON) network, which adheres to the convention of placing activities on (or at) nodes. This form of activity network is the one most often used in current project management software.

If we were to take a list of all our tasks, such as that generated by a very complete WBS, we could create a list or set of nodes, one for each activity. However, because we are interested in scheduling activities, it is absolutely essential that we properly determine the logical ordering, or precedence relationships, between them. In the simple construction example mentioned above, only a very foolish (or dangerous) construction project manager would expect to pour the concrete before the forms were constructed and in place. Similar relationships often exist in design projects. We would not, for example, attempt to determine which of a user's objectives are most important until we were sure that we had fully enumerated all of the objectives. Similarly, a team leader would encourage a design team to defer evaluating alternatives until after all of the alternatives have been generated. We thus see in these cases that the logical ordering of the project's activities will keep us from doing one activity until another is complete. This type of logical relationship is called *finish-to-start precedence*. That is, we must finish one activity prior to starting its successor activity.

Sometimes the logical relationships are more subtle. We might, for example, get a potentially useful design idea while we are still learning about the client's needs. In this case, the relationship is such that we can *begin* the task of generating ideas before we complete the task of understanding needs, but we can't consider the idea-generating task *complete* until we are quite sure that we have completed the task of understanding needs. This is called *finish-to-finish* precedence, and it means that we cannot finish the successor until we are sure that the prior activity is finished. One common example of finish-to-finish precedence occurs during the writing of a final report. Teams are sometimes encouraged to begin to write the final report of a project very early on, since that will allow them to get sections such as the literature review done while the literature is still fresh in their minds. However, it would be very disturbing to learn that a team had finished the final report before it had completed the design aspects of their project!

A third type of relationship is *start-to-start* precedence, in which one activity cannot be started until another one is started as well, although neither activity needs to be finished before the other. An example of this type of relationship might be editing sections of the final report for a project. In order to have a team member check grammar and style, some of the report must already have been written, but it doesn't have to be finished for checking to get started. (Clearly, in this and similar instances, judgment must be exercised. Checking that is done before report sections are substantially written may well have to be repeated as such minimally completed sections are subsequently further documented.) It is important to understand the different types of relationships between tasks, because if we otherwise assume that all activities are either fully independent or obey finish-to-start precedences, we are wasting opportunities to get work done earlier when resources may be more readily available or even otherwise idle.

Once we understand the logical relationships, we can draw an activity network for our project. Basically, this network consists of a node for each activity, and an arrow out of each activity to its logical successors. There is usually a number of activities that do not have any logical predecessors or any logical successors. At the start of a project, activities such as basic research or initial meetings with the client may not depend on anything before them; to deal with this, there is a convention that calls for starting an activity network with a "Start Project" node. In some cases, there may actually be a formal start, such as a project initiation meeting, but in others this is what is known as a *dummy activity*. Similarly, to ensure that every node has some place to connect its outward arrow, there is by convention an "End Project" node. Once again, there may be a formal ending to the project, such as turning in the final report or meeting with the client for the last time, but if not, the network should have a dummy node to end it. One consequence of the "Start Project" and "End Project" convention is that every node in the network except these two will have at least one arrow leading into it and at least one arrow leading out. If the activity has logical precedents, then the inward arrow should come from them. If the activity has logical successors, then the arrow(s) out of the activity should point to the successor.

We show in Figure 4.6 a simple activity network for the beverage container design project that indicates the logical relationships between this project's activities. Such a network can also take advantage of and incorporate the known duration of each activity. (Recall that in Section 4.3 we said that each task in the WBS should be broken down at least to a level such that we can estimate how long it will take to complete that task.) The activity network does not, however, fully schedule the tasks. The logical relationships in Figure 4.6 tell us that we cannot conduct the testing until we have both developed the test protocol and generated alternatives (which may include building mock-ups or prototypes). Note, however, that we can begin to develop the test protocol as soon as our research has been completed and we have met with the client's research director. If the research takes only a few weeks, and building a testable mock-up takes much longer, then we have a number of different times at which we could decide on a protocol. There is no simple formula for deciding when to decide; this decison calls for management judgment. If we think the team is going to be really busy during part of this time frame, and idle during another (such as when waiting for parts that have been ordered), we would likely want to plan on doing the protocol during that idle time. This illustration highlights an important notion, namely, that there are some activities whose start time we can adjust without affecting the overall time-to-completion of the project, while there may be other activities for which we should start as soon as is logically possible. An activity for which we can adjust a start time is said to have *slack*, which is the time difference between the earliest date we can start that activity, and the latest date on which we can start it without affecting the overall timing of the project. Activities that have no slack are said to be on the *critical path*. Starting and completing such critical path activities on time is of central importance to the timely completion of a project. That is why successful project managers pay very close attention to activities that are on the critical path, while staying "only" mindful of the others.

We certainly do not intend to fully explain all of the subtleties of this type of scheduling, but it is certainly worthwhile for a design team to determine which activities must be carefully kept on schedule, and which are off the critical path. Adjusting activities that are not on the critical path is also important for balancing the workload of the team. In fact, it is generally better to plan a team's work so that it can do things at a steady or relatively constant pace, since the normal occurrences of Murphy's Law will always create opportunities for things to pile up near the end. Therefore, it is helpful to plan the work and move activities that have

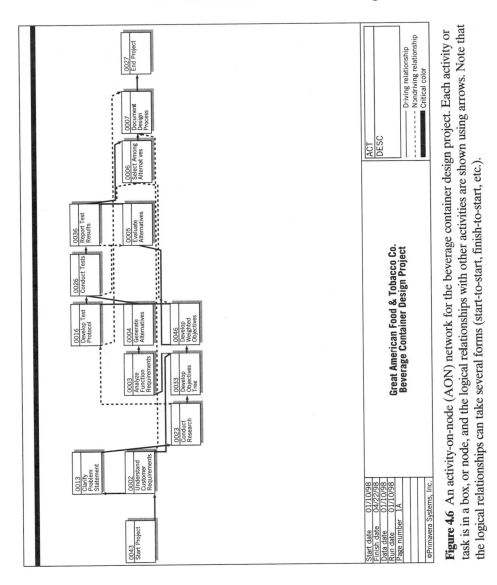

Figure 4.6 An activity-on-node (AON) network for the beverage container design project. Each activity or task is in a box, or node, and the logical relationships with other activities are shown using arrows. Note that the logical relationships can take several forms (start-to-start, finish-to-start, etc.).

some slack toward the beginning of the project. We sometimes say that in design projects, it is better to organize the work so that we can panic early, and then later events that might induce panic have already been dealt with. (We have to say that our experience suggests that if you panic early, you will still get the chance to panic later, but it likely will be about more interesting and useful problems.)

4.5.3   Gantt Charts: Making the Timeline Easy to Read

The same information that we incorporate into an activity network can also be represented as a bar graph, or *Gantt chart*. This approach to scheduling is said to have originated with Henry Gantt, one of the early founders of the field that is now known as industrial engineering. Gantt was charged with improving the output of munitions production during World War I, and he determined that it would be useful to formally track and diagram the processes involved. Apparently he was quite successful, as his method is widely used today not only to track processes, but to plan them.

Figure 4.7 shows the Gantt chart that corresponds to the activity network of Figure 4.6. Note that for each activity there is a start date and an end date that can be read from the time axis at the top of the chart. Many managers find that the Gantt chart is more readable than an activity network since it allows us to relate time in a rather evident manner. Further, we can develop or sketch a Gantt without any computer-based tools, using only graph paper, although we can also easily do them within standard spreadsheet programs. This simplicity can come at a price, however. A project team may construct a Gantt chart without carefully thinking through the logical relationships discussed above. If this happens, then several problems will likely ensue as the project unfolds. First, it becomes harder for the team to set priorities if the logical relationships among tasks are unclear. The team may, for example, work on the most interesting or most difficult task, even if others are on the critical path. In addition, in the absence of accurate information regarding the logical ordering of events, the team may find that a seemingly unimportant activity has more important successors that cannot be begun. Finally, if the inevitable slippage occurs on some tasks, all of the slack in an activity may be consumed, and this then makes that activity and its successors critical.

These sorts of problems can be avoided if the activity network is assumed to relate to a Gantt chart with finish-to-start precedence. Thus, teams should work as a group to first construct the activity network. Then, some or all of the team members can display these results in a Gantt chart or another, similar graphical format. This helps to ensure that the team has considered all the tasks and their relationships to each another, and it reinforces the planning already undertaken in developing the LRC.

4.6   BUDGETS: KEEPING TRACK OF THE MONEY

Budgets are essential but difficult tools for project managers. They permit teams to determine what financial and other resources are required, and to match these requirements to the available resources. Budgets also require teams to account

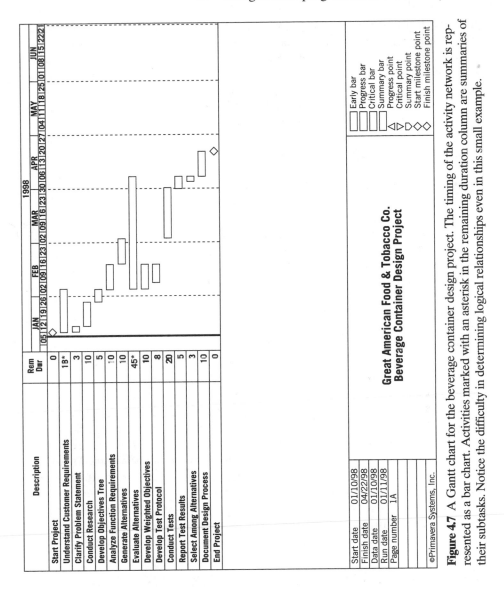

Description	Rem Dur	1998					
		JAN	FEB	MAR	APR	MAY	JUN
Start Project	0						
Understand Customer Requirements	18*						
Clarify Problem Statement	3						
Conduct Research	10						
Develop Objectives Tree	5						
Analyze Function Requirements	10						
Generate Alternatives	10						
Evaluate Alternatives	45*						
Develop Weighted Objectives	10						
Develop Test Protocol	8						
Conduct Tests	20						
Report Test Results	5						
Select Among Alternatives	3						
Document Design Process	10						
End Project	0						

Great American Food & Tobacco Co.
Beverage Container Design Project

Early bar
Progress bar
Critical bar
Summary bar
Progress point
Critical point
Start milestone point
Finish milestone point

Start date	01/10/98
Finish date	04/22/98
Data date	01/10/98
Run date	01/11/98
Page number	1A

©Primavera Systems, Inc.

Figure 4.7 A Gantt chart for the beverage container design project. The timing of the activity network is represented as a bar chart. Activities marked with an asterisk in the remaining duration column are summaries of their subtasks. Notice the difficulty in determining logical relationships even in this small example.

for how they are spending project monies. And, finally, budgets serve to formalize the support of the larger organization from which the team is drawn.

Many of the engineering economics concepts that we will discuss in Chapter 8 are relevant for budgets, so we will defer much of our discussion of these concepts at this time. Further, we usually don't need large, complex budgets for doing the sort of design projects that are likely to be done in academic or similar settings. (Remember, as we have said before, we are concerned with the budget for doing the designs, not with the budget for making the designed objects.) Thus, design project budgets normally include research expenses, materials for prototypes, and support expenses related to the project.

We will limit our budget discussions to the above categories of costs, that is materials, travel, and incidental expenses. This means that in attempting to budget for a design project, it is necessary at the outset to try to determine what sorts of solutions are *possible*. This is not to say that we have determined the solution, rather that we should consider resource needs earlier than might really be desirable. One effect of this is that design projects often try to establish *not to exceed* budgets that set limits on what can be expended by specifying the highest cost that might come about. The danger in this approach is that if it is done this way routinely, for all projects in an organization, resources will be set aside for all of the design and other projects that will never used.

As a final note, it is important to properly value the time invested in a design project by each and every member of a design team. This is important even in student design projects done in design courses. (In fact, there may be a tendency to undervalue this very scarce resource just because we did not call out that time in the budget.) One way to place a value on a team member's time is to adapt the "algorithm" that employers use to "bill out" the time of an engineer who is working on a project. Most firms charge between *two and four times* an employee's direct compensation when they bill a client for that employee's time. That multiplier covers fringe benefits, supervision, overhead costs, and profit. If we were to bill student time at a minimum wage rate of $6.00 per hour, a team of four students working ten hours each per week on a project for ten weeks would be billed out by a design firm at $4,800–9,600 for the entire project. Put in the simplest terms, time is a valuable, scarce, and irreplacable resource—don't waste it!

4.7   TOOLS FOR MONITORING AND CONTROLLING: MEASURING OUR PROGRESS

We have now, presumably, developed a plan, a schedule, and a budget. How can we track how our team performs relative to our plan? This is an important question, but it can be very hard to answer. If the plan for a construction project calls for a task to be completed by a certain date, the project manager can go out and see if the task has been done. If, for example, the plumbing is not in by the specified date, the project manager may not even have to turn on a faucet to see that there is a problem. In a design project, however, monitoring and control is more subtle and, in some ways, more difficult. Therefore, it is essential that the members of a team agree upon a process for monitoring their joint progress before the project gets very far underway.

There are a number of techniques and tools available to monitor projects, but these often involve team members filling out time sheets, punching clocks, or other accounting tools. For smaller design projects, especially academic projects, these kinds of tools may not be very effective. We will therefore focus on a simplified version of the percent-complete matrix. The *percent-complete matrix* (PCM) is widely used in the construction industry as a means of relating the extent of work done on the parts of a project to the staus of the overall project.

The goal of the PCM is to determine the overall status of the project. A PCM uses the information in the WBS and the budget. Constructing a PCM requires only that we know the cost of each item or area of interest, and the percent of the total cost corresponding to that item. Thus, the PCM allows us to input the percent of the work on that task or work item and, by summing over all the items in the project, we can calculate the total percent of the project completed. In general, the method is best suited to cases where some clear method of calculating progress is available. If, for example, the foundation work of a building project constitutes 25% of the total expected costs of a project, then we have completed at least 12.5% of the project when we have completed one-half of the foundation work. A project manager can periodically update progress in each of the general areas in the WBS to determine overall project progress.

In some cases a physical measure can serve as a proxy for progress, such as yards of concrete poured compared to the total volume called for in the plan, or tons of steel erected compared to budgeted totals. While this approach has significant appeal in standard projects, design projects are generally more concerned with progress relative to allowed time than to available budget, and physical measures are not likely to be available. One alternative is to use the estimated duration of the team's activities instead of dollar budgets, together with a simple rule for tracking progress. A simple rule is that if work on an activity is begun, then the team can immediately claim 33% (or 50%) progress for that activity. However, the team gains no additional progress on the task until the activity is completed. When completed, the team receives the remaining 67% (or 50%) credit for the task. In no case is the team given more than 100% credit for a task, no matter how long an activity really takes, and the team gets full credit when the work is done, regardless of how long it actually took. (This method clearly places a premium on careful and accurate decomposition of the work in the WBS.)

Consider Figure 4.8, in which we show a modified PCM for the beverage container design project. Each of the tasks used in the activity network has been included, except for the summary tasks, such as understanding customer requirements and evaluating alternatives. (The detailed breakouts for these have been included instead.) The PCM shows the planned or budgeted duration for each task, the percent of the total project that the task accounts for, and its status. In those cases where a task has either been started or completed, credit toward the overall project is given. In cases where there is no progress, no credit is given. Three observations are in order. First, the project manager or team leader could give a more exact percentage of completion than the 0, 33%, 100% used in this example (and in our simple rule), and could do so for selected tasks. Remember that it is the team that chooses the values in the simple, standard rule. Second, the team can compare

Figure 4.8 A percent-complete matrix (PCM) for the beverage container design project. Each activity and its share of the overall project is given. The team is given 33% credit when an activity is begun and the balance upon completion. Unless the tasks have been broken down sufficiently, this method can be misleading, but for small projects it does provide a reasonable approximation of progress.

Percent-Complete Matrix TASK	Planned Duration (days)	Percent of Total	Status (see key)	Credit (days)
Start Project	0	0%	2	0.0
Clarify Problem Statement	3	3%	2	3.0
Conduct Research	10	11%	2	10.0
Draft Objectives Tree	2	2%	2	2.0
Review OT	1	1%	2	1.0
Revise OT	2	2%	2	2.0
Analyze Functions	10	11%	1	3.3
Generate Alternatives	10	11%	1	3.3
Develop Weighted Objectives	10	11%	2	10.0
Develop Test Protocol	8	9%	1	2.6
Conduct Tests	20	21%	0	0.0
Report Test Results	5	5%	0	0.0
Select Among Alternatives	3	3%	0	0.0
Document Design Process	10	11%	0	0.0
End Project	0	0%	0	0.0
Total Days Budgeted	94	100%		39.6%

Key:
0 = Not Started, No Credit   1 = In Process, 1/3 Credit   2 = Completed, Full Credit

the progress achieved thus far with the overall time allocated for the project. If, for example, the design project were in the fourth week of a ten-week project, the PCM would seem to indicate that the project is more or less on plan. If the team was in the eighth week, this PCM would be cause for alarm. Third and last, we note that if the team has done a good job of determining the nature and duration of the tasks required to finish the project, this PCM and this method will allow them to monitor their work. If they have not, this method is simply an illusion.

4.8   MANAGING THE XELA-AID CHICKEN COOP PROJECT

As we have repeatedly noted, one of our illustrative examples is based on a student design project that was part of a one-semester, introductory engineering sign course given at Harvey Mudd College. Only a limited number of project management tools could be successfully learned and applied in this context. On the project management side, the student teams asked to design a chicken coop for the residents of San Martin Chiquito, Guatemala, were asked to develop and submit only WBSs. Two examples are given in Figures 4.9 and 4.10, and several comments are in order.

First, the two teams chose quite different approaches, with one team using a graphical presentation and the other a tabular chart. The WBS in Figure 4.9, aside from the team's oversight in not including lines, is quite similar to that in Figure 4.1. We can see that this WBS has several problems, the most obvious of which is that several of the subtasks are not actually breakdowns of the higher order tasks. Note that "Analyze Function Requirements" has an associated subtask, "Black Box Functional Analysis." Absent another subtask, the one given is simply a restatement of its parent task. Remember that the WBS is a way of breaking down the work into constituent components or smaller jobs.

A second problem with the example in Figure 4.9 is that it is not complete. The team will certainly have some additional management responsibilities, including team meetings and progress reports. The figure doesn't include them. An incomplete WBS will ensure that all other plans that depend on it will also be wrong.

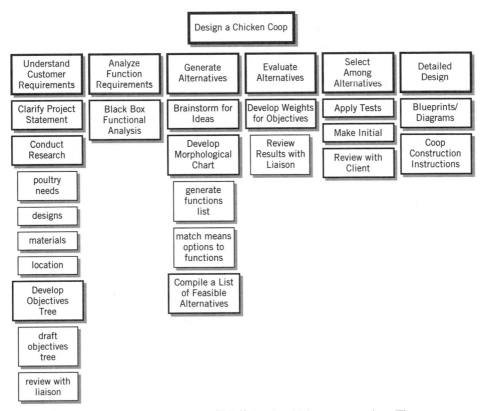

Figure 4.9 A work breakdown structure (WBS) for the chicken coop project. There are several errors in this WBS that are commonly made by less experienced project managers. Can you spot them?

Figure 4.10 A task list that could serve as the basis for a WBS for the chicken coop project. Note that the team has estimated the time requirements for the project by summing up over all the tasks. If placed in hierarchical order, this could be very easily adapted to a WBS.

	task	person hours
i.	talk to client	5
ii.	read client statement	1
iii.	create objectives tree	6
iv.	create task list	3
v.	revise client statement	7
vi.	create function box	4
vii.	talk with client about work above	2
viii.	revise objectives tree, function box, and task list	7
ix.	research Guatemala	7
x.	research chickens and coop designs	16
xi.	other research	4
xii.	create a morphological chart	6
xiii.	combine means on morph chart into coherent designs	4
xiv.	determine the best conceptual combination of means	6
xv.	discuss further alternatives	5
xvi.	discuss with client	2
xvii.	acquire materials for scale model	8
xviii.	build scale model	12
xix.	build segments of full-size replica	20
xx.	documentation of building process	5
xxi.	determine metrics for evaluating design	10
xxii.	evaluation of physical model as well as conceptual design	20
xxiii.	redesign portions of coop if necessary	10
xxiv.	rebuild these segments	10
xxv.	retest coop	5
xxvi.	evaluate coop	5
xxvii.	present design to client	2
xxviii.	make small changes if necessary	5
xxix.	prepare written report	20
xxx.	prepare how-to-build instructions	5
xxxi.	prepare oral presentation	20
xxxii.	present to class and client	2
	total time to complete our project	**245**

The task list in Figure 4.10 is more complete, even including the team's estimate of time required to complete each task. It goes so far as to include the time needed to develop the task list (item *iv*) The real problem with this table is that it is a list, not a hierarchy, and so it doesn't show the relationship between tasks. This may seem unimportant to the novice project manager, but understanding the relationships among tasks is an important insight when decisions must be made about the activities that have to be done first, and, under time pressure, those activities that can be left undone.

4.9   NOTES

Section 4.1: The definition of projects is from (Meredith and Mantel 1995).

Section 4.2: The underlying conceptual model for the 3S approach is from (Oberlander 1993). The example in Figure 4.2 is adapted from (Kezsbom, Schilling, and Edward 1989).

Section 4.3: The text-based WBS example shown in Figure 4.3 is from a sample set included in the software Primavera Project Planner, Release 2.0.

Section 4.4: Linear responsibility charts are explained further in most introductions to project management, for example, (Meredith and Mantel 1995).

Section 4.5: Activity networks and Gantt charts are covered in greater detail in any introductory text on project management, for example, (Meredith and Mantel 1995).

Section 4.7: The use of the standard form of the Percent Complete Matrix is given in (Oberlander 1993). The form as modified draws upon a method given in (CIIP 1986).

Section 4.8: The WBSs of Figures 4.9 and 4.10 are from (Gutierrez et al. 1997) and (Connor et al. 1997), respectively.

4.10   EXERCISES

4.1 Explain the differences between managing design projects and managing the implementations of design projects. For example, consider the differences between designing a highway interchange and building that interchange.

4.2 Develop a work breakdown structure (WBS) and a linear responsibility chart (LRC) for an on-campus benefit to raise money for the homeless.

4.3 Develop a work breakdown structure (WBS) and a linear responsibility chart (LRC) for an on-campus meeting to be used by the women in the cooperative cited in Exercise 3.5 as they arrive at a nearby airport and travel to your campus. (Hint: You are not allowed to do the courteous thing and pick them up at the airport.)

4.4 Develop a work breakdown structure (WBS) and a linear responsibility chart (LRC) for a project to design a robot that will be entered into a national collegiate competition.

4.5 Develop an activity network for the on-campus benefit of Exercise 4.2.

4.6 Develop an activity network for the robot project of Exercise 4.4.

4.7 Develop a schedule and a budget for the on-campus benefit of Exercise 4.2.

Chapter 5

Specifications

How can I express what the client wants in terms that help me as an engineer?

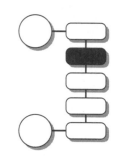

$\mathbf{U}$p to this point we have focused on understanding what the client and users want and need from a design. In this chapter we move from using the language of the client to the language of the engineer. In particular, we are interested in translating the client and user needs and desires into terminology that helps us find ways to realize those needs and measure how well we have met them. Expressed in the terminology of engineers and designers, we identify functions, performance specifications, and metrics for the artifact. These three terms represent distinct, albeit related, aspects of how a designed artifact actually does what it was designed to do.

Recall that in Chapter 1 we indicated that the *specifications* of artifacts are precise descriptions of the properties of the object being designed. They are often (but not always) expressible as numbers or measures. We also defined several types of specifications, including *design specifications* that articulate in a numerical or measurable way what a design is supposed to do and provide the basis for *evaluating* that design because those "specs" become the "targets" of the design process against which we measure our success in achieving them.

We also drew distinctions in Chapter 1 between prescriptive specifications, procedural specifications, and performance specifications. These are different ways of formalizing what the client or user wants in terms suitable for engineering analysis and design. *Prescriptive specifications* are used to specify values of attributes that the designed object must meet to be successful. For example, a chicken coop must be able to safely house 25 full grown chickens. *Procedural specifications* identify or mandate the specific procedures or methods to be used in calculating attributes or behavior. For example, the ability to house a chicken safely might be specified by requiring application of the USDA poultry guidelines. *Performance specifications*, which we will discuss in more detail below, characterize the desired behavior or performance of the designed object or system. Performance specifications have come to take on several different meanings in different contexts, so we will need to carefully consider what we mean when we use this term in given situations.

We will now look first at *functions* and functional specifications; these tell us *what* the thing must do in order to realize the stated objectives. Until we have clearly established which functions have to be performed, it is very difficult, if not impossible, to accurately specify how well the functions must be performed. We will then consider *performance specifications*; these "specs" tell us *how well* the thing must do something. Finally, we will introduce *metrics*; these are the tools for testing and measuring the performance of a design in meeting a specification or an objective. Our ordering of functions, specifications, and metrics is somewhat arbitrary. For example, in many cases a designer will begin to consider how the realization of a particular objective might be measured before she has determined precisely the function that is being performed. In other cases, some or even most of the performance specifications may have already been set out by the client at the beginning of the design job. It is important to remember (and understand!) that we are presenting these design tools and techniques in a particular order because it delineates one of the ways of thinking about them and it follows a reasonably logical sequence. This is, however, a case where the iterative nature of design that we discussed in Chapter 2 will almost surely lead to some variation in the timing of when functions, performance specifications, and metrics are both introduced and processed.

5.1   FUNCTIONAL SPECIFICATIONS: WHAT FUNCTIONS MUST BE PERFORMED TO REALIZE THE OBJECTIVES?

If we were to ask a child what a bookcase does, he would probably answer that "It doesn't *do* anything, it just sits there." An engineer, however, would note that the bookcase does a number of different things, and that it must do them well to be a successful design. For example, the bookcase resists the force of gravity exactly, so that books neither fall to the floor nor are forced into the air. The shelves of the bookcase may have dividers, to aid in separating the books into categories chosen by the owner. If the bookcase has been designed with an eye to interior design, we might say as designers that it enhances the visual appeal of the room. Each of these ways that our designed object, the bookcase, does things, are functions. So, when an engineer sees things that were designed, she has trained her eye to see that artifacts do things, even when they "just sit there."

It is essential that we understand what a designed object must do if we are to create a successful design. In this section we will first look more deeply at what we mean when we talk about a design doing something. Then we will look at techniques for understanding and listing functions.

It is particularly important for a designer to be able to properly specify functions. There are consequences for the engineer who fails to understand and design for *all* of the functions in a design. The literature of forensic engineering is rich with cases in which engineers failed to realize some additional function(s) that should have been met, often with tragic results.

5.1.1   What Are Functions?

We can think of functions in a number of different ways. In elementary calculus, for example, we write that $y = f(x)$ and we say that y, a dependent variable, is a *function* of x, a single independent variable. That is, the value of y depends on the value of x. In multivariate calculus we extend this notion to include multiple independent variables and multiple dependent responses. Management studies also refer to *transformation functions* in which we say that a vector of inputs (labor, materials, technology, etc.) is transformed into a set of outputs (products, services, etc.). In both of these cases we are highlighting the existence of a relationship between some independent variables (i.e., *inputs*), and response or dependent variables (i.e., *outputs*) and we are characterizing that relationship in a formal way. We use a similar, although not identical, notion of functions when we consider the functions of an object we are designing. To designers, functions are simply the things that the designed object must do in order to be successful. As such, the statement of a function usually consists of a verb and a noun, and the verb will normally be an "action" verb, rather than a "being" verb. For example, lift, raise, move, transfer, or light are all action verbs. The noun in the statement of function will generally start off as a very specific referant or item, but an experienced designer will look for the more general case.

For example, one of the functions of the bookcase we discussed was to resist forces of gravity, which we might have characterized as "support books." However, this would have implied that the bookcase would be used to hold only books. But shelves in bookcases often support trophies, art, or even piles of homework. Thus, a more basic—and more useful—statement of the function to be served in this case is that shelves will resist the force of gravity associated with any object weighing less than some predetermined weight. That is, our statement of function is that shelves will support some predetermined number of kg (or lbs). When describing functions, then, we should use a verb-noun combination that best describes the most general case.

We also want to avoid tying a function to a particular solution. If we were designing a cigarette lighter, for example, we might be tempted to consider "applying flame to tobacco" as a function. This implies that the only way to light the tobacco is by using a flame (and that tobacco is the only material to be lit). Car lighters, however, use electrical resistance in a wire to achieve this function. Thus, a better way to characterize this function might be "ignite leafy matter," or even "ignite flammable materials." (And, somewhat parenthetically, we consider the following questions. In light of the well-documented health hazards associated with smoking, is there an ethical issue for an engineer who is asked to design a "better" cigarette lighter? Is this is an appropriate design task? We will discuss ethics in engineering and design in Chapter 9 but we do note that the issue does not arise if we considered the lighter as a tool for use in camping or survival training.)

We can also categorize functions as being either *basic* or *secondary* functions. Here a *basic function* would be defined as "the specific work that a project, process, or procedure is designed to accomplish." *Secondary functions* would be (1) any other functions needed to do the basic function or (2) those that result from doing the basic function. Secondary functions can themselves be either required or unwanted functions. *Required secondary functions* are clearly those needed for the basic function. Consider, for example, an overhead projector. Its basic function is to project images, and it has required secondary functions that include converting energy, generating light, and focusing images. *Unwanted secondary functions* are undesirable byproducts of other (basic or secondary) functions. For the overhead projector these could include generating heat and generating noise. These undesirable byproducts often generate new required functions, such as limiting the noise generated or dissipating the heat generated.

5.1.2   How Do We Identify and Specify Functions?

Now that we know what functions are, we turn our attention to the issue of identifying and specifying the functions to be performed by the artifact that is the focus of our particular design project. Our goal in determining these functions is to ensure that our final design does everything that it is supposed to do. Several methods have been developed for determining functions, and in this section we describe four: enumeration, analysis of "black" and "transparent" boxes, construction of function–means trees, and reverse engineering, a method that is also known as dissection.

5.1.2.1   *Enumeration*

Perhaps the most basic (and most obvious) method of determining functions for a designed object is to simply *enumerate* or list those functions that we can readily identify. This is an excellent way to begin functional analysis for many objects. It leads us to consider what the basic function of the object is and it may prove useful as well in determining secondary functions. In many cases, however, we could get "stumped" very early in this process. Consider, for example, a bridge. If the bridge is used for highway traffic, we might note that its basic function is to act as a conduit for cars and trucks. Most of us would have to scratch our heads for a while before being able to go much further than this initial, single-entry list. However, there are a few useful "tricks" we can use to help us extend a list we are enumerating.

One trick is to imagine that an object exists and then ask what would happen if some or all of it suddenly vanished. If our bridge disappeared entirely, for example, any cars on the bridge would fall into the river or ravine over which the bridge crosses. This suggests that one function of the bridge is to support any loads placed on it. If the abutments ceased to exist, the deck and superstructure of the bridge would also fall, which suggests that another function of the bridge is to support its own weight. (This may seem silly until we recall that there have been more than a few disasters in which bridges collapsed because they failed to support even their own weight during their construction. Among the most famous of infelicitous bridges is the Quebec Bridge over the St. Lawrence River, which collapsed once in 1907 with the loss of 75 lives and again in 1916 when its closing span fell down.) If the ends of a bridge that connect to various roads disappeared, traffic would not be able to get on the bridge (and any vehicles on the bridge would be unable to get off). This suggests that another function of a bridge is to connect a crossing to the road network. (We will explore this function in some depth in a case study in Chapter 10.) If road dividers on our bridge were to disappear, vehicles headed in one direction might well collide with vehicles headed in the other direction. Thus, separating traffic by direction is a function that many bridges serve, and it is a function that can be accomplished in a number of ways. For example, the George Washington Bridge in New York City assigns different directions of traffic to each of its two levels. Other bridges use median strips.

Another useful approach for determining functions is to consider how an object might be used and maintained over its lifetime. In the case of our bridge, for example, we might note that it is likely to be painted, so that one function is to provide the bridge's maintenance workers with access to all parts of the structure. This function might be served with ladders, catwalks, elevators, and so on.

Consider once again our beverage container design problem. Here, because we have ample experience with such containers, we can readily name or list the functions served by a beverage container, including at least the following:

- contain liquid
- get liquid into the container (fill the container)
- get liquid out of the container (empty the container)
- close container after opening (if container is to be used for more than one time)
- resist forces induced by temperature extremes
- resist forces induced by handling in transit
- identify the product

Note that the functions of getting liquid into and out of the container are distinct. This is evident after a brief reflection on canned beverages: Liquid is sealed in by a permanent top, while access is obtained through a pull tab. We might have noticed this distinction between the filling and emptying functions had we considered the "life cycle" of a beverage container. Thus, one of the "tricks" that we have just mentioned might also have been applicable here.

We also observe that at the heart of our approaches to function enumeration lies the need for the designer to think about a designed object and to list the verb-noun pair that corresponds to each and every function. However, because enumeration alone is often difficult, we must turn to other methods.

5.1.2.2   Black Boxes and Transparent Boxes

Recall that our earlier discussions of mathematical and management functions featured inputs and outputs, both singly or in groups. This input-output model is also often useful for thinking about engineered systems and their associated functions. One tool that helps us relate such inputs and outputs, as well to the transformations between them, is the *black box*. Essentially, a black box is a graphic representation of the system or object being designed, with the inputs shown entering the box on its left-hand side and the outputs leaving on the right-hand side. It is important that all of the known inputs and outputs be specified, even those undesirable byproducts that result from unwanted secondary functions. In fact, in many cases such functional analysis can help us identify additional inputs or outputs that may have been overlooked. Once the black box has been initially drawn, the designer can ask questions such as "What happens to this input?" or "Where does this output come from?" We can answer such questions by removing the cover of the black box, thus making it into a *transparent box*, so we can see what is going inside. That is, the point of making a box transparent is to expose transformations from inputs to outputs. Within a given box, especially after it's made transparent, we can also link more detailed "subinputs" to internal boxes that produce related "suboutputs."

Consider, for example, a system that has three inputs: an airborne signal within that part of the frequency spectrum containing the radio frequencies (RF), a controllable source of electrical power, and a vector of desired outputs (such as particular stations and volume levels). There are three obvious outputs for this system: sound, heat, and (almost surely) a display that indicates how the user's desired frequency and volume level were realized. This system, known to all of us as the radio, can be thought of as being contained in a box that transforms the RF signal into another signal that we hear as audible sound, that is, music, talk, and perhaps noise! How does this happen? What functions are performed in a radio? Can we identify the many functions of our radio by looking inside the box that contains our radio?

We can model the functions of our radio with the series of boxes depicted in Figure 5.1. In Figure 5.1(a) we see the radio's most basic function, namely, the conversion of an RF signal to an audio signal, is achieved, while at the same time several byproducts, both desired and undesired, are generated. If we take the cover off this box (Figure 5.1(b)), we see several new black boxes within. These

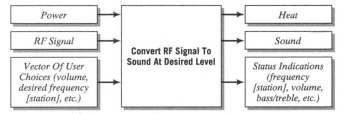

Figure 5.1 (a) This is a black box for the radio. Notice that all the inputs and outputs are somehow related in the single, top-level function. We call this top-level function a *basic function.* If we want to know how the inputs are actually transformed into the outputs, we need to remove the cover on the black box.

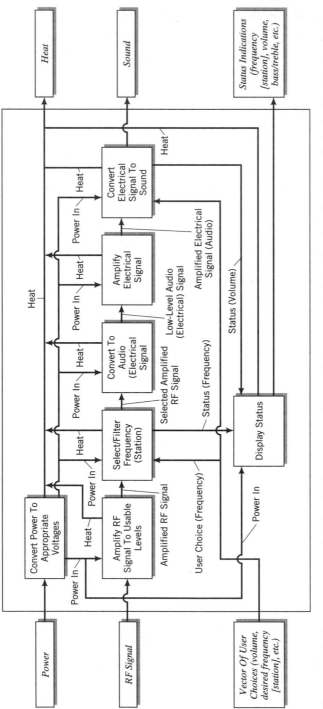

Figure 5.1 (b) The cover on the black box has been removed, or made transparent. Notice that in order to transform the inputs into outputs, a large number of secondary functions are required. If our design responsibilities (or our curiosity) demanded it, we could also remove the covers on some or all of these functions. Notice also that this design calls for the heat to be allowed out of the box on its own. In many designs, we would add a specific function, "dissipate heat," and then decide on a strategy for doing so.

boxes include transforming the power from that of a wall outlet, 110V, to a level appropriate for the radio's internal circuitry, probably 12V. Other internal functions include filtering out unwanted frequencies, amplification of the signal, and conversion of the RF to an electrical signal that can drive speakers. Thus, making our black box cover transparent revealed a number of additional functions. If we were responsible for designing the radio, we would probably want to remove the covers of even more of the boxes we now see. If, on the other hand, we are simply assembling a radio from known parts, we might stop at this level. This method of making internal boxes transparent and analyzing their internal functions is also called the *glass box method*. No matter which term we use, the effect is the same: We keep opening internal boxes until we fully understand how all inputs are transformed into corresponding outputs and what additional inputs or side effects are produced by these transformations.

The black box method can be a very effective way of determining functions. However, it would be an unfortunate mistake to think that it applies only to devices that have a physical box or housing. The only requirement for using a black box/transparent box is that the designer is willing and able to identify *all* of the inputs and outputs. If, for example, we are designing a playground to be used in a rainy climate, our inputs would include the children, their parents or caregivers, and the rain. Our outputs would include entertained children, satisfied parents, and water. If we forget the water, our playground design will neglect the need for proper drainage. (As an aside, it is generally not sufficient to include general terms such as "weather" unless we are willing to consider how the weather is translated into water, ice, wind, and heat *within* our box.)

A final point on the transparent or glass box method is that there is a serious need to carefully define the *boundaries* of the device or system for which we are identifying functions. We face a potential tradeoff when we set such limits. If the boundaries are set too widely, we may be setting functions that are beyond our control, for example, if we wanted to specify the household electric current in our radio example. On the other hand, if boundaries are set too narrowly, they may serve to limit the scope of the design. In the case of the radio, for example, it may be an open issue whether the radio output is an electrical signal that is fed to the speakers or whether the radio output is the acoustical signal coming from the speakers. The question being decided by the placement of the boundary, then, is whether or not the speakers are part of the system. Such decisions are generally resolved by the client and potential users.

5.1.2.3   *Function–Means Tree*

Sometimes we have some ideas about how our designed device or system might work from the earliest stages of the design process. While we have counseled against "marrying your first design" and cautioned against trying to solve design problems before they are fully understood, it is often true that our early design ideas suggest very different functional aspects. Consider again our hand-held (cigarette) lighter. Clearly, if we use a flame to ignite leafy materials we will encounter some differences in the secondary functions than if we use hot wires, or, for that matter, lasers. One such difference could be the shielding of the igniting element if the device is small enough to be pocket-borne by the user. In those cases where considering means or implementations can lead us to different functions, a function–means tree can help us sort out secondary functions.

A *function–means tree* is a graphical representation of the basic and secondary functions associated with a design. The top level of the tree shows the basic function to be met. Each of the succeeding levels alternates between showing various means by which the primary function might be implemented and displaying the secondary functions made necessary by those means. In order to make the tree more readable, some graphical notation is employed to distinguish functions from means; for example, functions and means can be shown in boxes with different shapes or written in different fonts. Figure 5.2 shows part of a function–means tree for the hand-held cigarette lighter. Note that the top-level function has been specified in the most general terms possible. At the second level, two different means have been given: a flame-based lighter and a hot-wire-based lighter. These different means imply a different set of secondary functions, as well as some common ones. Some of these secondary and common functions and their possible means are given in lower levels.

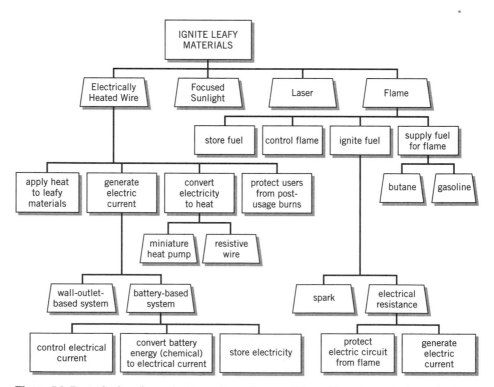

Figure 5.2 Part of a function–means tree for a cigarette lighter. (Functions are shown in rectangles, while means are shown in trapezoids.) Notice that there are different subfunctions that result from different means. It is often the case that conceptual design choices will result in very different functions at the preliminary and detailed design stages.

Once a function–means tree has been developed, we can list all of the functions that have been listed, noting which are common to all (or many) of the alternatives and which are particular to a specific means. Those functions that are common to all the means are likely to be inherent in the problem. Others will need to be addressed only if the associated concept is adopted during the alternatives evaluation phase (which we will discuss in Chapter 6).

A function–means tree has another useful property in that it begins the process of associating what we must do with how we might do it. In Chapter 6 we will return to this issue when we present a tool that helps us generate and analyze alternatives. That tool, the morphological chart, lists the functions of the designed artifact and the possible means for realizing each function in a matrix format. The effort we put into the function–means tree really pays off then.

Several cautions should be noted regarding function–means trees. The first, and probably most obvious, is that the use of a function–means tree should not serve as a substitute for either the problem formulation stages or the alternatives generation tasks that we take up in Chapter 6. There is always a temptation to use the outcome of the function–means tree as a complete description of the available alternatives. Doing so will likely restrict our design space much more than should be the case. A second caution comes in terms of using function–means trees without using some of the other tools described above. One commonly-made mistake that novices (or students) make is that they adopt a tool because it somehow "fits" with their preconceived ideas for a solution. This transforms the design process from a creative, goal-oriented activity into simply a mechanism for making choices that a designer wanted. That is, because the function–means tree allows us to work with means or implementations that may seem appealing, there is a tendency to overlook some functions that might have been revealed by a less "solution-oriented" technique.

5.1.2.4 *Dissection and Reverse Engineering*

Most engineers, and indeed, most curious people, ask the question "What does this do?" when confronted with a button, knob, or dial. A natural follow-up may take the forms "How does it do that?" or "Why would you want to do that?" When we follow up these questions with remarks on how we might do it better or differently, we are engaging in the art of *dissection* or *reverse engineering*. Reverse engineering means taking an artifact or device that does some or all of what we want our design to do and dissecting—or deconstructing or disassembling—it to find out, in great detail, just how it functions or works. We may be restricted from using that design for any number of reasons: it may not do all the things we want, or do them very well; it may be too expensive; it may be protected by a patent; or it may be our competitor's design. But even if all of these reasons apply, we often can gain insight into our own design problem by looking at how other people have thought about the same or similar problems.

The process is actually quite simple. We begin with a means that has been used by a previous designer, then determine what functions are realized by that given means. We can then begin to consider alternative ways of doing the same thing. If, for example, we wanted to understand the functioning of an overhead transparency projector, we might find a button which, when depressed, turns on the projector. One of the functions that the projector button has is to control when the projector is turned on or not. This can be done by a number of means, including push buttons, toggle switches, or bars along the front of the projector. It is an interesting exercise to consider just how many functions can be thought of for this commonplace device.

Several cautions should be noted about using reverse engineering in order to determine functions. First, the devices or objects that are being dissected were developed to meet the goals and objectives of a particular client and a target set of users. These people may, collectively, have had quite different concerns than are called for in the current project. This means that a designer should exercise great care to remain focused on the current client's needs. Second, there is often a temptation to limit the new means to those that work in the context of the object that is being reverse engineered. In the case of the classroom projector, for example, all of the means for turning on and off the power to the device are more or less compatible with a stand-alone device. In some settings, however, it might be more appropriate to remove these controls from the device itself and make them part of some more general room controls. In theaters, for example, we often find that the lights and other controls are managed from the projection room, rather than by wall switches such as those found in most houses or classrooms. It is important that we do not become captive to the design being used to assist our thinking.

A third point of interest is that while we treat the terms dissection and reverse engineering as equals, they may not always refer to exactly the same process. This is because dissection is sometimes viewed just as it is in a high school biology laboratory, wherein we dissect a frog, for example, so that we can see and learn about the frog's underlying anatomical structure. In this sense, dissection is more descriptive than analytical. In reverse engineering we go a step further as we try to determine means for making functions happen, which means that we are trying to analyze both the functional behavior of a device and how that functional behavior is implemented.

There is a fourth consideration, one that we stated earlier but need to reiterate here. We need to define functions in the broadest possible terms and only focus down when it is necessary. Restricting functions to the most immediate terms found on the object being reverse engineered can lead us into forcing our design to mimic someone else's, rather than fully appreciating any opportunities for new ideas. Finally, there are clearly some serious intellectual property and ethical issues tied into reverse engineering. We must always bear in mind that it is inappropriate to claim as our own the ideas of others. In some cases this can also be a violation of law. We will discuss intellectual property and ethics in Chapters 6 and 8, but it is important to appreciate the importance of honoring the ideas of others at least as stringently as any other (tangible) property they might hold. After all, wouldn't we want the same protections for our own ideas?

5.1.3   A Repeated Caution About Functions and Objectives

The sets of tools used to determine objectives and to determine functions are sufficiently similar that they are often confused by novice or fledgling designers, so that lists of objectives are often being made when functions are called for, and vice versa. This confusion appears to arise for two reasons: two new concepts—objectives and functions—are being learned, and the tools themselves appear to be very similar. The "newness" problem can probably be overcome only by trying to list examples of the two concepts and by talking about them with team members, teachers, and experienced designers, hoping to obtain in this way a practical feel for both objectives and functions. The confusion related to the tools themselves, however, can be helped by keeping in mind whether the immediate focus is on "being" terms or on "doing" verbs. As we noted in Section 3.1.2, objectives tell us what we want the designed artifact to be like, that is, what the final object will "be," and what qualities it will have. As such, objectives detail attributes and are usually be characterized by present participles such as "are" and "be." Functions, on the other hand, tell us what the object will "do," with a particular focus on the physical transformations that the artifact or system will accomplish. As such, functions often relate inputs to outputs, and are usually characterized by active verbs. Ultimately, this is a terribly important distinction, but its centrality is often only grasped fully as a result of a great deal of serious practice.

5.2   PERFORMANCE SPECIFICATIONS: CAN WE EXPRESS THESE FUNCTIONS IN TERMS OF THINGS WE CAN MEASURE?

In the previous section we looked at methods of determining what our designed object or system must do. While this is essential to our design process, it is difficult to use such functional specifications without also considering the question of *how well* the design must do them. If, for example, we want a design that can produce musical sounds, we need to be able to specify how loudly, how clearly, and with what tones the device must produce the sounds. For our purposes, we will refer to these performance specifications as either *prescriptive performance specifications*, or, alternatively, *definition-based performance specifications*. In addition, if our system or artifact is required to work with other systems or artifacts, then it is essential that we specify the ways that the various systems are to interact. We will refer to these particular performance specifications as *interface performance specifications*. Finally, it must be noted that when a design is completed, its end users will want to know what levels of performance they can reasonably expect from the designed object. We will refer to these (predicted) performance levels as *detailed design performance specifications*. It is perhaps regrettable that all three of these clearly distinct intellectual objects have come to be known as performance specifications, since using the same term to capture three ideas has the effect of confusing the meanings of all three ideas. Nonetheless, it is unfortunately true that many practicing designers and their clients use performance specifications to mean all of these things, and more. We would caution a clarity of thought about these matters, for the most obvious of reasons.

5.2.1   Putting Numbers on Prescriptive Performance Specifications

It is normally up to the engineer or designer to convert objectives and functions into terms that allow the application of scientific principles in general, and the engineering or applied sciences in particular, to the design problem at hand. One element of applying the engineering sciences that is essential for developing and assessing designs is the ability to translate functions into measurable terms. We must find a way to measure the performance of a design in realizing a specific function or objective and then establish the range over which that measure is relevant to the design. Finally, as designers, we must determine the extent to which improvements in performance over that range really matter.

Determining the range over which a measure is relevant to a design and deciding how much improvement is worthwhile are interesting problems. In Figure 5.3 we show a curve that is analogous to something that economists call a *utility* plot. It is our conceptual starting point for analyzing alternative designs for a device so that we can assess the marginal value of producing a gain in performance at a cost of some undesirable quality. The utility or value of the design gain is plotted on the vertical axis or ordinate over a normalized range from 0 to 1. The level or "cost" of the attribute being measured is shown on the abscissa or horizontal axis. By way of example, consider using processor speed as a parameter or measure for designing a laptop computer. At processor speeds below 100 MHz, the computer is so slow for modern applications that trying to improve from, say, 75 MHz to 80 MHz, provides no real gain. Thus, for processor speeds below 100 MHz, the utility is 0. At the other end of the utility curve, say, above 500 MHz, the tasks for which a computer is typically used can't exploit gains in processor speed. For example, word processing or browsing the Net may be more constrained by typing speed or communcation line speeds, so that a gain from 550 MHz to 575 MHz still leaves us with a normalized utility of 1. Thus, we've *saturated* our utility plot at high speeds, which is to say that there's no gain to be made by improving a design to make it faster within a range that is already very desirable.

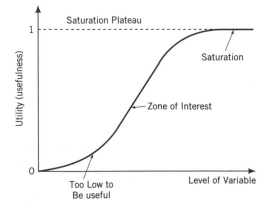

Figure 5.3 A hypothetical performance specification curve. Notice that until some minimal level is realized, no meaningful benefit is achieved. Similarly, above some saturation plateau, there is no meaningful benefit in gaining still more. The actual shape of the curve is likely to be uncertain in most cases.

However, what happens between, say, 100 MHz and 500 MHz? In this range we expect that changes do matter, and that improvements in processor speed serve to improve our incremental or marginal gain. We have sketched in Figure 5.3 an *S-* or saturation curve that indicates *qualitatively* what we think is happening. There are clearly gains to be made as we move toward faster speeds, but at some points these incremental gains decrease to small, marginal improvements as we approach the plateau at saturation. Thus, for the entire S-curve, the utility is initially flat (or 0) at low processor speeds, increases over a range of interest, and finally plateaus at 1 because added speed is not usable.

This sort of behavior depicted in Figure 5.3 is rather common. Economists refer to the law of diminishing returns, and even children with unlimited chocolate eventually stop eating because they get stomachaches! In our case, however, this isn't a "law," strictly speaking. We don't usually know the actual shape or precise details of the S-curve (it may not look nearly as smooth as what we have sketched in Figure 5.3), so we choose to approximate it by a collection of straight lines, such as that shown in Figure 5.4. Here there are still regions where gains no longer interest us, as indicated by the horizontal lines at levels 0 and 1. In the middle range, however, we suppose that we are just as happy to increase our levels of the design variable (e.g., processor speed) to obtain a corresponding linear gain in the utility, anywhere within this range of interest. Perhaps most important here is the point that, qualitatively, we are simply saying that the straight line defines the range within which we expect to achieve design gains by tuning the design variable in question.

This entire exercise may seem a little artificial, so let us consider another example. Suppose a designer is asked to develop a Braille printer that is clearly quieter than competing designs. Suppose further that the printer is to be used in office settings, and that none of the competing designs are quiet enough to be so used. A natural and important question to ask is, "How quiet does this design really have to be?" In order to answer this question, the designer must determine the relevant units of noise measurement and the range of values of these units that are of interest. The designer would also find out how much noise is generated by current printer designs and whether or not listeners can distinguish between different

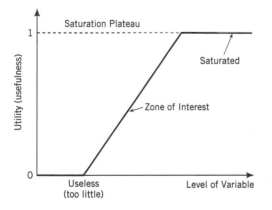

Figure 5.4 A linear approximation of the hypothetical performance specification curve shown in Figure 5.3. In this case, the design team has agreed on the lower (minimum) and upper (saturation) levels of interest, and is assuming that an equal increase anywhere along the sloping line brings the user an equal gain.

designs. If, for example, one printer produces the same noise level made by a pin dropping on a carpet, while another generates noise at the level of a ticking watch, it is unlikely that the designer will expend much effort to distinguish between them. Both are so quiet as to be fully acceptable. Similarly, if one printer design is as loud as a gas lawn mower, while another is as loud as an unmuffled truck, it is unlikely that either design will be useful in an office setting, so that there is no utility gained by distinguishing between these two designs. (Note, by the way, that this example is of a *reverse* S-curve in which we start at saturation because there is no gain to be made at such low levels of quietness and that we degrade to a zero level of utility for printers that are all uniformly too loud.)

Sound intensity levels are usually measured in decibels (dB), so a designer might conclude immediately that some range of dB is likely to be of interest. Carrying this further, the designer might look for some indication of how much noise is produced by other devices and within different environments. We show such sound intensities for various devices and environments in Table 5.1. For reference, we show the noise exposure levels to which workers may be exposed in Table 5.2. These levels, expressed in hours of exposure, are defined by OSHA, the federal agency concerned with the safety of work environments (remember that we referenced OSHA requirements while designing a safe ladder in Chapter 3). With such environmental and exposure information in hand, the designer can identify a range of interest for a performance specification for a Braille printer. New printer designs must make less than 60 dB noise in an office environment. Further, lesser values of generated noise levels are considered gains, down to a level of 20 dB. All designs that generate less than 20 dB are equally good. All designs that produce more than 60 dB are unacceptable. (And we note that any realistic designs will generate noise levels that are so far below the OSHA exposure values that occupational safety is not an issue here.)

Table 5.1 Sound intensity levels that are produced by various devices and are measured in various environments. Sound intensity levels are measured in decibels (dB) and are a logarithmic expression of the square of acoustic power. Thus, a 3 dB shift corresponds to a doubling of the energy produced by the source, while the human ear cannot distinguish between levels that differ by only 1 dB (or less).

Level (dB)	Qualitative Description	Source/Environment
10	Very Faint	Hearing Threshold; Anechoic Chamber
20	Very Faint	Whisper; Empty Theater
30	Faint	Quiet Conversation
40	Faint	Normal Private Office
50	Moderate	Normal Office Background Noise
60	Moderate	Normal Private Conversation
70	Loud	Radio; Normal Street Noise
80	Loud	Electric Razor; Noisy Office
90	Very Loud	Band; Unmuffled Truck
100	Very Loud	Lawn Mower (Gas); Boiler Factory

After (Glover 1993).

Table 5.2 Permissible noise exposures in American work environments, expressed in intensity levels (dB) permitted during various daily durations (hours). These levels and durations are defined by the Occupational Safety and Health Administration (OSHA). If workers are exposed to levels above these or for times longer than indicated, they must be given hearing protection devices.

Daily Duration (Hours)	Sound Level (dB)
0.5	110
1	105
2	100
3	97
4	95
8	90

After (Glover 1993).

5.2.2 Setting Prescriptive Performance Specifications

We can extend the above discussion to define a process for setting performance specifications. First, we determine design parameters that reflect the functions or attributes that must be measured and the units in which those parameters are to be measured. We then establish the range of interest for each design parameter. For desirable design variables (i.e., qualities or attributes), utility values below a threshold are treated as equals because no meaningful gains can be made. Utility values above a plateau or saturation level are again indistinguishable because no useful or worthwhile gains can be achieved. (We are assuming a standard S-curve in which the threshold comes first and the plateau last.) The range or zone of interest lies between the threshold and the plateau. It is within this zone that the design gains should be matched to and measured with respect to the design parameters that are the subject of a given prescriptive performance specification. This process works well when we exercise judgment in setting performance specifications based on: sound engineering principles, an understanding of what can and cannot be reasonably measured, and an accurate reflection of both client's and users' interests.

Consider once again our beverage container. Each of the functions that were specified earlier (in Section 5.1.2.1) and many of the objectives already given (in Section 3.1.4) may have a range of values we need to specify. Some of the functions and some relevant questions with each associated function are:

- contain liquid: How much liquid must the container contain, at what temperatures? Is there a range of fluid amounts that we can put into a container and still meet our objectives?
- resist forces induced by temperature extremes: What temperature ranges are relevant? How might we measure the forces created by thermal stresses on the container designs?
- resist forces induced by handling in transit: What are the range of forces that a container might be subject to during routine handling? To what degree should these forces be resisted in order for the container to be acceptable?

Note that similar but distinct problems arise for the second and third functions on this list, as they both relate to forces.

We can now develop a set of performance specifications that our container designs should meet by addressing these and similar questions. For example, we might indicate that each container must hold 12 ± 0.01 oz. In this case the specification has really become a constraint because the corresponding utility function is a simple binary switch: We either meet this design specification or we don't. (Of course, it is possible to study the container design problem as one in which a variable single-serving size is possible, in which case there may exist a linearized S-curve for the container size where smaller is better.) Still another performance specification might be that the containers can be filled by machines at a rate from 60–120 containers per minute. In this case, any container that cannot be filled at least this fast creates a production problem, while a faster rate might exceed current demand projections. We might specify that the designs should allow the filled containers to remain undamaged over temperature ranges from –20 to +140°F. Temperatures lower than a threshold of –20°F are unlikely to be encountered in normal shipping, and temperatures higher than a plateau of +140°F indicate a storage problem. In this case, it might turn out that some designs that appeal in other ways are limited by either temperature extreme. Then a judgment will have to be made about the importance of this objective and its associated performance specification. We discuss the analysis of the relative importance of objectives and their role in selecting a design alternative in Chapter 6.

5.2.3   Interface Performance Specifications

As previously noted, some designers also use performance specifications to specify how devices or systems must work together with other systems. These specifications, known as *interface performance specifications*, are particularly important in cases where several teams of designers are working concurrently on different parts of a final product, and smooth working of all of the parts is required. One example might be a designer of automobile radios who must ensure that the final design is compatible with the space, available power, and wiring harness of the car. Another example might be a case where the design team has agreed to divide the project up into several parts; the team must make sure that the final parts will fit together. A team that is developing a gate might separate the design of the mechanical elements that lift the gate from the design of electronic elements that control when the gate goes up or down or how it reacts to an unusual condition. In each of these cases, the boundaries between the systems must be clearly delineated, and anything that crosses the boundaries must be specified in sufficient detail to allow all teams to proceed.

Understanding and working with interface performance specifications is an increasingly important issue for large industrial and commercial firms that, in an increasingly competitive international arena, are trying to minimize the total time needed to design, test, build, and bring to market new products. Most of the world's major automobile companies, for example, have been able to reduce their design and development times for new cars to one-half or less of what they were a decade ago. In order to do this, design teams work concurrently on many systems or products, all of which must work together and be suitable for manufacture. This puts a premium on the ability to understand and work with interface performance specifications.

The process of developing interface performance specifications is conceptually quite simple, although the practice can be extremely difficult. At the conceptual level, the designers must first specify boundaries or interfaces between the systems that must work together and then develop a set of specifications for each item that crosses a boundary. These specifications might include a range of values (e.g., 5 V, ± 2 V), or some logical or physical devices that enable the boundary crossing (pinouts, for example, or even physical connectors), or simply an agreement that the boundary must never be breached (e.g., between heating systems and fuel systems). In each case, however, the designers for systems on each side of a boundary must have reached a clear agreement about where the boundary is and how it is to be crossed, if at all. In practice, this process can be difficult and demanding since teams on all sides are, in effect, placing constraints on the others. One idea that may be particularly helpful in working through interface specification development is black box functional analysis. This allows all the parties to identify the inputs and outputs that must be matched and to deal with any side effects or undesired outputs.

It is beyond our scope to cover this topic in greater detail, but it is worth noting that an immediate grasp of the issues involved in interface design can be gained by considering a simple example, such as the design and installation of a new toilet for a building. There are a number of interfaces or boundaries to be considered, including: connecting to the water supply to ensure that fresh water can enter the system, connecting to the sewer system to ensure that outputs are removed, and user interfaces. Note that each of the interfaces must be carefully specified and agreed to or the pipes will not connect, the toilet will not fit into the space provided, or the user will not be willing to sit down.

5.2.4 Detailed Design Performance Specifications

We should also mention the specification of the performance of a device in the context of how that device is used *after* it has been designed and manufactured. Clearly, as designers we hope that our design will meet or exceed every single prescriptive performance specification, and that the interfaces our design offers to users and consumers are appropriate for its intended use. End users, however, are usually not parties to the design process, and so they depend on published *detailed design performance specifications* that set out the performance levels that can be expected from a device or system. And while such performance specs are very important to the end user, they are not seen as particularly useful to the designer, especially during conceptual design. However, detailed design performance specifications can be quite valuable when they are used to identify questions or issues of performance that are likely to matter to potential end users or consumers. Designers can, in many instances, examine the performance specifications of other similar or competing designs and gain insight into those issues that ultimately do affect end users.

Consider, for example, the case of an electronic device that must work with other devices. Examination of the detailed design performance specifications of these other devices will almost certainly lead a designer to recognize constraints in the form of existing interface performance specifications. For example, if we are purchasing a computer, whether laptop or desktop, we will surely want to know whether or not it can work with whatever data and communication lines we have available, as well as whether or not it can support the software or applications that we intend to run, and at levels of speed and compatibility that we want. That is, we may be happy to run a simple drawing package locally, or we may have to do complex computer graphics calculations on large files maintained on a nonlocal server. In either case we will want to know what we are getting from our computer before we buy and install it in our office.

5.3 METRICS: HOW DO WE TEST VARIOUS ALTERNATIVES?

Having determined what our client would consider a *good design* (objectives), what the design *must do* (functions), as well as *how well it should do* at least some of these things (performance specifications), we now take up the question of how we might determine how well a particular design *actually does* all these things. We answer this question by introducing *metrics*, that is, standards of measurement. In this context we will use metrics to measure the extent to which a design realizes our objectives. Ideally, a metric gives us an exact gauge of the objective we are concerned with. In practice, we often make difficult choices about what constitutes an appropriate metric, how we can actually apply that metric, and how much it costs to measure the achievement of a design objective.

5.3.1 Steps for Developing Metrics

We follow a three-step procedure in selecting metrics: (1) identify both the units and the scale of something that it is appropriate to *measure* about our objective; (2) identify a means of assessing the value of a design in terms of those relevant units; and (3) evaluate whether this particular measurement and its subsequent evaluation is feasible. In many cases we will be forced to cycle through this process iteratively until we come up with a suitable method of measuring the design in terms of the objective. The process is described in more detail below.

Our first step in establishing metrics for a design attribute is to determine the appropriate units that might be applied to the objective we are concerned with. Recall our earlier example of a portable computer. For our objective of low weight, we might consider units related to weight or mass, that is, kg, lb, or oz. For the objective of low cost, we would probably want our metric to be the cost measured in currency, that is, $US in the United States. The appropriate "units" for some cases may be general categories, or even subjective rankings (i.e., "high," "medium," and "low"). While these are inherently unappealing to most engineers and scientists, they are often quite adequate for assessing conceptual design alternatives.

What sorts of scales are available? In the context of product design, for example, there are six (at least) types of measurement scales that have been proposed as appropriate for testing and evaluating designs. These six scales are shown in Table 5.3. The important thing to note about these different types of scales and their associated units of measure is that they can each be used in different situations, but with an awareness of the limits and restrictions that each implies. Nominal scales, for example, can be useful when distinguishing among categories, but cannot be subjected to statistical tests without some form of summing up.

Once we have established appropriate units of measure and scale, we must take the second step of determining how to accurately assign a figure or value to a particular design. An important aspect of this step is to assure that the plan for measuring the design's performance is compatible with the type of scale and measure selected in the first step. This could include, for example, laboratory tests, field trials, consumer responses to surveys, focus groups, etc. Given an objective of low weight, we could determine the weight by using a conventional balance scale. Cost, on the other hand, could be quite difficult to measure, unless we know factors such as the manufacturing techniques to be employed, the number of units to be manufactured, and the components to be included in the design. Estimating costs can be a complex and demanding field (that we will return to in Chapter 8). For now, however, let us assume that we can estimate the costs to manufacture and distribute our portable computers.

Our third step in assigning metrics is the determination of whether or not the information derived from using a metric is worth the cost of actually performing a measurement. In some cases we will find that the usefulness of the metric is slight when compared to our own resources or to those needed to obtain the measurement. In such cases we can either develop a new metric, find another means for measuring the expensive metric, or look for an alternative way of assessing our design. There may be several metrics available with which we are equally comfortable, in which case we may be able to select a less expensive alternative. In other cases there may be less expensive means for obtaining a

Table 5.3 Measuring scales for testing and evaluating designs in the field of product design.

Nominal scales, such as colors, smells, or even professions
  (e.g., teachers, lawyers, engineers)

Partially ordered scales, such as grandparent, parent, uncle, and child, which array
  themselves somewhat in order of seniority

Ordinal or *rank order* scales, such as first, second, third, etc.

Interval scales, such as degrees centigrade

Ratio scales, such as inches, seconds, or dollars
  (Unlike interval scales, ratio scales have a true zero.)

Multidimensional scales or *index numbers,* such as miles per gallon or kilometers per
  maintenance event, that are compounds of other scales of measurement

---

Adapted from (Jones 1992).

value of the metric. In still other cases, we may decide to use a less accurate method to assess our designs. As a last resort, we may decide to convert our objective into a constraint, which allows us to consider some designs and reject others. (Remember that constraints allow us to exclude options from the design space, while objectives allow us to choose among alternative designs.)

Consider again our objective that our laptop computer design be low cost. It may be that the information needed to accurately assess the manufacturing costs for portable computers is not available to the design team without a significant and expensive study. An alternative option might be to estimate the manufacturing cost by adding up the costs of the individual components when purchased in given lot sizes. This disregards a number of relevant costs (e.g., assembling components, overhead), but it would allow the design team to distinguish between designs with expensive elements and designs with cheaper elements. Alternatively, the designers might depend upon expert input from the client and simply rank the designs into ordinal categories such as "very expensive," "expensive," "moderately expensive," "inexpensive" or "cheap," and "very cheap." Failing all else, the designers might be forced to reconstruct this into a constraint, such as, "contains no parts costing greater than $200."

5.3.2   Characteristics of Good Metrics

Good metrics have a number of attributes. The first of these, of course, is that the metric should *actually measure the objective* that the design is supposed to meet. Often we find that designers try to measure some phenomenon that, while interesting, is not really on point for the desired objective. If, for example, our objective is to appeal to consumers, then measuring the number of colors on the package may be a poor metric. A second characteristic is that the metric should be capable of the *correct level of accuracy* or *tolerance*. If "low weight" is one of our objectives, measuring to the nearest ton or to the nearest milligram is not appropriate, at least not for a laptop computer.

A third characteristic of concern is that the metric should be *repeatable*. That is, if others were to conduct the same test or measurement, they would obtain the same results, subject to some degree of random error. This characteristic can be met either by using standard methods and instruments, or, if no such methods are available, by carefully documenting the protocols being followed. It also makes it incumbent upon the design team to use sufficiently large statistical samples where possible. A fourth characteristic of good metrics, which is related to repeatability, is that the outcomes should be expressed in *understandable units of measure*, as we have discussed in the previous section.

Finally, any metric should elicit only *unambiguous interpretation*. That is, we would like to have metrics whose results lead all of the members of the design team (and all other stakeholders) to the same conclusion. We certainly don't want to be in the position of arguing among ourselves about the meaning of a measurement of a given metric.

It is clear that judgment is called for in selecting a good metric. The scale and units should be appropriate to the design objectives, and means of measuring them must be available and affordable. In general, good metrics result from careful thought, extensive research, and ample experience—which suggests that the selection of metrics can certainly be enhanced by the synergy derived from a cooperative, well-functioning team.

5.3.3   On Metrics and Performance Specifications

The distinction between metrics and performance specifications is very subtle, and often a source of confusion. Both, for example, can involve quantifying or otherwise specifying how well a design does something. In some cases, metrics that are adopted for an objective may be the same as those applied to functions. Nevertheless, there are important differences that we should keep in mind.

The most important difference is that *metrics apply to objectives (only)*. Metrics allow both designers and clients to assess the extent to which an objective has been realized by a particular design. Performance specifications, on the other hand, are typically applied to functions and specify how well those functions must be realized by a design. In this sense, performance specifications take on something of the characteristics of constraints—any design that fails to meet a performance specification may be considered to have failed. Metrics are needed for *all* of the objectives that are being considered in the design selection process; performance specifications apply only to those functions for which there are definable limits of acceptability.

5.4   METRICS AND SPECIFICATIONS FOR THE XELA-AID CHICKEN COOP DESIGN

In the previous chapter, we clarified and documented the objectives for the Xela-Aid chicken coop design project and refined the problem statement. In this section we will present some examples of functional analysis, performance specification, and metrics that the student teams developed. As we have noted earlier, our purpose in presenting student design work is not to criticize them or to find fault with their work. Rather, such presentation serves two other ends. One is that it demonstrates the largely successful application of the principles and techniques we are describing. The second end is that the ability to review design documents is another important design skill. It is rather unlikely that any design team will be both complete and accurate in their first versions of the tools we've presented—which means that team members have to be able to evaluate their initial efforts with patience and prudence, to further the design process.

A chicken coop actually has a number of functions that can be associated with it. One of the design teams used the enumeration method to develop the list of functions shown in Table 5.4. Notice that many of the functions take the form of "allow. . ." While this is a reasonable way to begin the process of functional analysis, it is often the case that a function can be cast in a more active form. For example, "allow for removal of waste" might be better stated as "remove waste." Similarly, "allow for chicken feeding" can be more succinctly stated as "feed chickens." Other functions listed in Table 5.4 suggest solutions, for example,

Table 5.4 One list of functions for the Xela-Aid chicken coop, as developed by one of the freshman design teams.

Protect chickens from predators during the day
Protect chickens from predators during the night
Protect chickens from weather
Allow for feeding chickens
Allow for watering chickens
Keep water fresh
Allow for egg collection
Protect eggs
Allow incubation of eggs
Ventilate coop
Allow for nest cleaning
Allow for chicken entry
Allow for waste removal
Allow human entry/exit from coop
Allow human entry/exit from perimeter

"keep water fresh" seems to imply that the water should be changed frequently. There are, of course, alternative solutions, including removing old water, streaming water continuously, and detoxifying the water. It should be clear that a more general statement of function might be "supply chickens with potable water." Now, this discussion may seem like nitpicking, but it is important that we, as designers, make it a habit to specify functions in general terms that do not imply solutions. If we don't do it even for relatively simple problems, we can expect to find great difficulties in more complex design problems where solutions are far less obvious.

In Figure 5.5 we show the black box analysis of a chicken coop done by another student team. The tables below the diagram list some of the inputs and outputs and their roles in various aspects of coop use. Such tables can help designers as they begin identifying what must go into and what must come out of a black box. The black box in Figure 5.5 has a number of interesting features and a few problems. Notice that the team has determined that "spoilt water"—which is an object, not a true function—is an output of the water container. This allows the team to identify both a function, "separate spoilt water," and an output, "bad water." There are several functions that have no obvious outputs, such as sleep and exercise, suggesting that the design team has confused the functions of the coop with the functions of the coop's residents. Further review of this figure will no doubt highlight other problems and errors in this example.

(The teams responsible for developing the chicken coop designs were not required to develop performance specifications for their designs. While this leaves a hole in the documentation, it also leaves an interesting mental challenge or thought exercise. For example, what are some of the prescriptive design specifications that could apply? How might they be derived? What is the interface between a chicken coop and the surrounding area?)

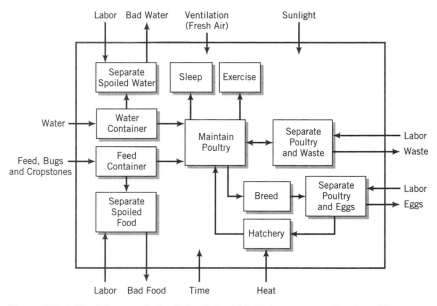

Figure 5.5 A black-box analysis of the Xela-Aid chicken coop as developed by one of the student teams. There are several problems with this black box. One is that some of the boxes have no outputs. This seems unlikely, and should raise questions immediately. More fundamentally, it is not clear if this is a black box for the chickens or for the coop.

The teams did develop metrics for their objectives. In the two cases presented here, the teams quickly realized that measuring the performance of designs against some of the objectives was simply beyond the scope of a one-semester, introductory design project. Figure 5.6 shows some of the metrics that one of the teams developed. While these metrics show an interesting grasp of the underlying concerns (e.g., the cigarette test for preventing the spread of fire), most of them are beyond the time and other limits of the team. As a result, this team essentially abandoned formal metrics for most of the objectives and instead based its decision on their own estimates of the number of chickens that each coop design could sustain. This represents an extreme—and not recommended—version of the iteration process that was discussed above. Figure 5.7 shows an example of simpler metrics and associated tests that another team developed and used. Essentially, this team elected to test individual components used within their overall design rather than focus on the design itself. Notice that the tests are reduced to simple Yes or No outcomes from simple field tests. This has the virtue of being quite manageable. Unfortunately, it also has the effect of separating the tests from their underlying objectives, thereby making it unclear exactly what the point of each test might be. This approach also leads the team to the rapid adoption of a very small set of design alternatives, within which the selection of components remained somewhat larger. It can be argued that in both cases, the teams might have been more successful had they gone through several thoughtful iterations, as was proposed in Section 5.3.2.

Figure 5.6 This table shows some of the tests (metrics) that were proposed by a student design team for the chicken coop. Many of the tests appear to require either more time or more resources than the team actually had. In this case, the team would have been better served by simplifying the metrics or the tests.

survive floods	simulate flood conditions by using a fire hose on a proto-type to see if the footings can be washed away or if the roof leaks
withstand temperature changes	attempt to destroy the materials used in the design through repeated freezing and heating (within the constraints of Guatemalan climate)
easy feeding and easy cleaning	ask a volunteer to do the tasks involved and measure the time and effort expended
inexpensive	determine the ration of net cost to the number of chickens maintained by the coop
reproducible	build a scale model, then give it (and it alone) to a volunteer and ask her to build another one like it out of the same materials
easy access for caregivers	test for presence, absence of tight (less than three feet wide) corridors or high (over five feet) accesses
prevent the spread of fire	test a model's resistance to various types of fires, such as airborne sparks (on the roof), cigarettes dropped in nesting material, etc.
appropriate tools and materials	whether or not they are locally available
protect from predators	build a prototype, put bait inside cages and nest boxes, and then see if any predators can get to it
prevent egg cannibalism	presence of dark nests (based on authoritative references)
increase production	the number of cages and nest boxes available in the design
promote bodily protein growth	credit if chickens have access to the ground to scratch for insects
maintain suitable temperature (for chicks and eggs)	full credit if broody hens are used; if artificial heat is used, test variability of heat source by placing a thermometer near it and measuring the variance of temperature
inhibit spread of disease	make a full prototype with live chickens; infect one chicken with a disease, and track the spread through the population. If a significant number of chickens survive, the design passes.

Figure 5.7 This table presents an alternative approach to Figure 5.6. Because of limitations of both time and resources, the team has chosen to evaluate various components in their design, and has converted them to pass/fail tests. Are there similar tests that could be used for the objectives shown in Figure 5.6?

Coop component tested	Materials used in testing	Metrics	Did the component pass the test?
Rebar perimeter fence posts, individual	Rebar posts	Did it support weight when upright?	No.
Bound rebar fence posts	Rebar posts	Did it support weight when upright?	Yes.
Removable tray supports	Rebar	Did it support weight when horizontal?	Yes.
Gutter	Corrugated metal and gutter components	Would the corrugated roof support a gutter?	Yes, using only certain gutter hangers.
Ease of construction and "constructability"	Styrofoam, balsa wood, glue	Were the building instructions clear and complete enough to construct a model?	Eventually.

5.5 MANAGING THE SPECIFICATIONS STAGE

For many inexperienced designers, the specification stage is the most difficult. It requires practice and effort to learn to think in terms of functions and to regard them as intellectual objects distinct from objectives and constraints. To this end, team-based approaches can be quite helpful.

There are some engineering tasks that are better done by individuals (e.g., calculating a stress in a bridge cable), while there are others that are best done by groups (e.g., designing a major suspension bridge). Setting specifications is a task that has both individual and group components. Many teams find it useful to have members attempt to use the methods individually initially, and then to have the entire team review, discuss, and revise these individual results at a later time. This sort of divide-and-conquer approach ensures that team members are prepared for meetings and that there is an experienced practitioner for each of the tools or methods. Group review and revision allows team members to build on each other's ideas in developing and setting functional specifications, so that the teams benefit from fresh viewpoints and further critical thinking.

Setting up metrics is another area where a team will usually do a better job than its individuals, *if* the members come prepared. One of the natural outcomes of the initial work by individual members is that metrics are developed that are not practical for the team. Team members are usually all too willing to highlight the problems they see in each other's metrics. This not only serves to prevent the team from attempting tests or measures that are beyond their abilities, or challenging them to expand and demonstrate such abilities, it can also lead to increased tensions. Team leaders and responsible team members should regularly try to keep the group focused on constructive, idea-based criticism, and away from destructive, personality-based criticism.

5.6 NOTES

Section 5.1: Further details about the Quebec Bridge and other "engineer's dreams" can be found in Petroski's history of American bridge building (1995). The black box for the radio used in Section 5.1.1 was developed by our colleague, Carl Baumgaertner, for use in Mudd's freshman design course. The term glass box method was coined in (Jones 1992). The function–means tree used was developed by our Harvey Mudd colleague, James Rosenberg, to illustrate an example originally proposed in (Akiyama 1991).

Section 5.3: The discussion of scales of measurement draws heavily from (Jones 1992).

Section 5.4: The results from the Xela-Aid chicken coop design project continue to be taken from final reports (Gutierrez et al. 1997) and (Connor et al. 1997).

5.7 EXERCISES

5.1 Explain the differences between functions and objectives.

5.2 Explain the differences between metrics and performance specifications.

5.3 Develop a set of metrics for the portable electric guitar of Exercise 3.2. If a metric is likely to be difficult to measure, indicate how it can be reframed as a constraint.

5.4 Using each of the methods for developing functions described in Section 5.1, develop a list of the functions of the portable electric guitar of Exercise 3.2. How effective was each of these methods in developing the specific functions?

5.5 Develop a set of metrics for the rain forest project of Exercise 3.5. If a metric is likely to be difficult to measure, indicate how it can be reframed as a constraint.

5.6 Using each of the methods for developing functions described in Section 5.1, develop a list of the functions of the rain forest project of Exercise 3.5. How effective was each of these methods in developing the specific functions?

5.7 Based on the results of either Exercise 5.4 or 5.6, discuss the relationship between methods of determining functions, the nature of the functions being determined, and the nature of the artifact being designed. Is the designer's level of experience also likely to affect the outcome of functional analysis?

5.8 Do the research necessary to determine whether there are any applicable standards (e.g., safety standards, performance standards, interface standards) for the design of the portable electric guitar of Exercise 3.2.

5.9 Describe the interfaces between the portable electrical guitar of Exercise 3.2 and, repectively, the user and the environment. How do these interfaces constrain the design?

5.10 Developing countries often have different safety (and other) standards than are typically found in countries such as Canada and the United States. How could this affect the design of both the portable electric guitar of Exercise 3.2 and the rain forest project of Exercise 3.5?

Chapter 6

Finding Answers to the Problem

Which of these ten good ideas is the best answer?

In Chapter 3 we described the problem definition phase of the design process and in Chapter 5, where we delineated design specifications, we started our discussion of conceptual design. Now we are ready to finish our articulation of conceptual design and begin to explore preliminary design. Here we want to generate concepts, or schemes or embodiments, of designs that will meet our objectives. To accomplish that we want to (1) generate candidate or potential designs, (2) organize them in ways that make it easy to explore them, and (3) evaluate them to see which are worth pursuing and which are not. In so doing, we will also discuss the building of proof-of-concept models and prototypes.

6.1 WHAT IS A DESIGN SPACE?

A design space is a simple, yet broad concept. A *design space* is a framework, a mental construct of an intellectual space that envelops or incorporates all of the potential solutions to a design problem. Because it is such a broad concept, the notion of a design space has only a limited amount of utility, and that lies in its ability to convey a *feel* for the design problem at hand. We invoke the image of a *large design space* when we are talking about a design problem for which (1) the number of potential designs is very large, perhaps even infinitely so, or (2) there is a large number of design variables and they, in turn, can take on a large number of values.

Two interesting examples of design problems having large design spaces are passenger aircraft (e.g., the Boeing 747), and large commercial office buildings (e.g., Chicago's Sears Tower). These are two engineered objects that have a lot of parts. There are six million different parts on a 747, and one can only imagine just how many parts there are in a 100-story building, ranging from structural rivets to window frames to elevator buttons. With so many parts, there are surely even more choices, even more design variables. Thus, these two artifacts have very large design spaces. On the other hand, these projects differ from one another because airplanes are constrained in ways that buildings are not, just as their performance presents a different challenge. In fact, the architect and structural engineer designing a high-rise building may have innumerably more choices for the shape, footprint, and structural configuration of their skyscraper than do aircraft designers when they are designing fuselages and wings. While weight is important as the number of floors and occupants of a building mount up, and while the shapes of high-rise buildings are analyzed and tested for their response to wind, they are certainly subject to far fewer constraints than are the payload and aerodynamic shape of aircraft.

As we will discuss further in Section 6.2.2, constraints play an important role in limiting the size of a design space, which makes it interesting to ponder what happens in the absence of any, or very many, meaningful constraints. For example, suppose we were chartered to design a road or transportation network for a brand new city. (Lest this be viewed as a fanciful notion, new suburban tracts and cities are being created all the time. Some of the more famous of such endeavors are Levittown, New York; Columbia, Maryland; Reston, Virginia; and Celebration, Florida.) We might imagine a designer being given, effectively, a blank sheet of paper and asked to lay out one or more maps of brand-new streets. Perhaps the external world imposes some constraints, such as ramps to nearby interstate highways or stops on nearby rail or commuter lines. But the design space of new boulevards, roads, and culs-de-sac would be very large unless a large number of meaningful constraints is set forth. Without constraints that impose some requirements or ordering on a design, the space of opportunities can be virtually limitless.

We refer to a *small design space*, or one that is *bounded*, when we discuss a design problem for which (1) the number of potential or candidate designs is very limited, or (2) there is a small number of design variables and they, in turn, can take on values only within limited ranges. The design of individual components or subsystems of large systems often takes place within relatively small design spaces. For example, the design of windows in both aircraft and buildings is sufficiently constrained by opening sizes and materials that their design spaces are comparatively small in size. Similarly, the range of framing patterns for low-rise industrial warehouse buildings is limited, as are the kinds of structural members and connections used to make up those structural frames.

6.1.1 Complex Design Spaces and Decomposition

Design spaces are certainly complex when they are large, if only because of the combinatorial possibilities that emerge when hundreds or even thousands of design variables must be assigned. However, design spaces can be complex because of interactions between various subsystems and components, even if the number of choices is not by itself overwhelming. For example, we mentioned in Section 2.2.5 the challenge of designing paper-handling systems, such as paper transports or paper feeders, to move paper through a photocopier. The design of paper transport systems is done within a complex design space. First of all, a paper transport must meet a large set of requirements, including some on the geometry of the subsystem to which paper is being fed (e.g., where the paper enters and exits), and some on the properties of the paper on which copies are made (e.g., its size, weight, stiffness, and other mechanical properties). Further, constraints arise from the need to move paper rapidly, precisely, and crisply, so there are very tight tolerances on the related engineering parameters.

In addition, the design has to accommodate conflicting constraints in order to handle different sizes of paper. Paper is moved along a path in the copying machine by having a set of devices that "grab" and move individual sheets. One constraint is to insist that no more than two such devices grab any individual sheet at one time, while another is that every sheet must be grabbed at least once. Thus, prescribing or choosing a separation distance between adjacent sets of grabbing devices means balancing these two conflicting constraints while trying to accommodate copied sheets of varying lengths or sizes. A related complex design issue is the choice of grabbing device and its parametric design in the face of a wide range of papers.

It is clear from even this cursory discussion that designing paper-handling systems involves making decisions about: geometry, spatial layout, timing, the forces involved in grabbing paper and in keeping sheets from getting jammed as they whisk around corners, making it possible to easily clear out jammed paper, and much more. In fact, one aspect of design complexity is that collaboration with many specialists is essential because it would be quite rare for a single engineer to know enough to make all of the design choices and analyses.

Again, this complexity emerges because values of many design variables are highly dependent either on choices already made or on those yet to be made. Beyond acknowledging their complexity, how do we attack such complex design spaces? Our approach is to apply the concept of *decomposition*, or, as it is sometimes called, *divide and conquer*. Decomposition is the process of dividing or breaking down a complex problem into subproblems that are more readily conquered or solved. Designs of paper transports can usually be decomposed into subproblems: lay out a smooth path between the input and output locations; decide the number and location of grabbing devices (e.g., pinch roll pairs) to be placed along that path; design a "baffle" or container to enclose the paper path and guide the paper; design the individual components (e.g., the sizes of driver and idler pinch rolls); select the proper materials for components such as the pinch rolls and baffle; and so on. In other words, break down both the design problem and the design space into manageable pieces, and then take them on one at a time!

The tools that we present in the remainder of this chapter are themselves designed to assist in the process of decomposing our design problem into solvable subproblems and reassembling their solutions into coherent, feasible designs. The morphological chart described in Section 6.3 is particularly suited to (1) decomposing the overall functionality of a design into its constituent subfunctions; (2) identifying the means for achieving each of those functions; and (3) enabling the composition, or *re*-composition, of candidate solutions. The recomposition or synthesis of feasible or workable solutions is particularly important. Imagine how we'd feel, having taken apart a very elaborate clock mechanism in order to fix something, and then finding we couldn't put it back together because a part was just a bit too large or a fitting had a slightly different configuration! Similarly, when we are recomposing candidate designs, we have to be sure to exclude incompatible alternatives. That's why we present in Section 6.4 some ideas and tools for ensuring that our newly assembled designs can work.

6.2   EXPANDING AND LIMITING THE SIZE OF THE DESIGN SPACE

Having established the flavor of or feeling for what we mean by a design space, we now get closer to a fundamental issue in design: How do we generate design ideas? Or, in terms of our current metaphor, how can we usefully expand the size and range of our design space? To answer these questions we could launch a philosophical discussion about creativity, touching as well upon a potentially contentious argument about whether creativity can be taught and learned, or whether it is innate and genetic, as are the colors of our eyes and hair. But that would be well beyond our scope and, from a very pragmatic, engineering point of view, not very helpful. Instead, we will consider and answer these questions in terms of enhancing and supporting creativity. Thus we ask, What means and tools are available to help us generate design ideas and to enlarge and organize the space of potential designs?

However, before we answer this question, there are two brief points that we want to make about creativity in engineering (and we will amplify both along the way in the balance of this section, as well as in Section 6.7). The first is that engineering creativity is *goal-directed*, that is, it is creativity designed to serve a (known) purpose, not to search for one. The goal may be imposed externally, as is often the case in engineering and design firms, or internally, as in the new product idea being developed in a garage. But, *there is a goal* toward which the creative activity is aimed. The second point is that *creative activity involves work*. Thomas Edison said it best: "Invention is 99 percent perspiration and one percent inspiration." In other words, we will almost certainly fail if we are not prepared to do some serious work aimed at generating design alternatives.

6.2.1   Sources of Ideas for Expanding the Design Space

There are perhaps two central themes to our presentation of ideas for expanding the design space by generating alternatives. The first is that we rarely gain any advantage by reinventing the wheel. In other words, and particularly when we are defining functions and the means to execute them, we should be aware that it is likely that other people have already tried to implement some, many, or even all of the functions we may need to make happen. It is only common sense to suggest that we should identify, study, and then use the formidable amount of existing information that is already available. Thus, it should not be a surprise that much of what we will discuss in Sections 6.2.1.1 and 6.2.1.2 has parallels with our prior discussion of means for acquiring information (cf. Section 2.3.3.1).

The second principal idea is that the kinds and styles of team interactions that we will describe below have much in common with brainstorming. That is, we will also hark back to some of the ideas we advanced in Section 3.5.1 about brainstorming as we describe some activities that a design team, perhaps augmented by the addition of other stakeholders or experts, could undertake as its members work to generate design alternatives. Here we will suggest both particular activities a group can undertake (in Section 6.2.1.3) and ideas about things the members of a team might think about (cf. Section 6.1.2.4).

6.2.1.1 Taking Advantage of Design Information That is Already Available

In Section 2.3.3.1 we detailed the importance of conducting literature reviews to determine the state of the art and identify prior work in the field. This included locating and studying previous solutions, product advertising, vendor literature, as well as handbooks, compendia of material properties, design and legal codes, etc. *The Thomas Register* is one valuable digest of product vendors. It lists more than one million manufacturers of the kinds of systems and components that are most useful in mechanical design. The 23 volumes of the *Register* are found in most technical libraries, and they are updated every year. Further, while more and more material is becoming available on the world-wide web, there is far more information that is not on the web—and perhaps never will be. Thus, while web-surfing is a very useful form of information gathering, it should never be viewed as the only way to identify and retrieve design-related knowledge.

There are two information-gathering tasks that are central to product design. These tasks are (1) benchmarking competitive products to evaluate how *well* such existing products perform certain functions, and (2) reverse engineering competitive products to see *how* functions are performed and to perhaps identify better ways of performing similar functions.

Recall, too, that we discussed using surveys, questionnaires, and focus groups to find out how users responded to various design ideas. And we described the utility of informal interviews for better defining the design problem, and of structured interviews of experts in the domain of the design problem.

We repeat these means here to raise and drive home a point. One way to generate design alternatives and concepts is to revisit the notes that we took as the original design goals were being clarified. It is quite likely that design ideas and problem solutions emerged during the problem definition phase of the design process. Thus, looking back at our "old" notes allows us to recapture old or premature ideas and solutions that we might have recorded then.

6.2.1.2   Patents: Expanding the Design Space Without Reinventing the Wheel

One related activity that we can undertake while generating alternatives is to search for relevant patents that have been awarded. We would do this in part to avoid "reinventing the wheel," in part to leverage our thinking and build on what we already know about a still-emerging design. We might also do a patent search to identify available technology that we can use in our design, assuming we can negotiate appropriate licensing agreements with the patent holder(s).

Patents are a kind of *intellectual property*, by which we mean that holders of patents are identified as those who are given the credit for having discovered or invented a device or some way of doing things. In engineering practice, for example, an individual's list of patent awards has as much or more weight than corresponding lists of publications associated with academia. This intellectual credit is awarded by the U.S. Patent Office (USPTO) to individuals and/or to corporations after they file an application that details what they believe to be the *new art* or originality of their invention or discovery. Patent applications are subject to lengthy and thorough examinations by USPTO patent examiners, after which patents are awarded or denied. Since patents result from (examined) *claims*, they can be—and often are—challenged because others feel that they have developed the relevant *prior art* that forms the foundation of that challenged patent. From the designer's viewpoint, filing patent applications is a mixed blessing. The award of a patent does provide some protection for new and innovative ideas. At the same time, however, such patents may inhibit the development of ideas for second- and third-generation improvements of existing devices or processes.

There are two basic kinds of patents granted by the USPTO. *Design patents* are granted on the form or appearance or "look and feel" of an idea. Design patents clearly relate to an object's visual appearance, as a result of which they are considered relatively weak patents because only minor alterations in the appearance of a device are enough to create a new product. *Utility patents* are granted for functions, that is, on how to do something or make something happen. Utility patents are considered stronger and and often cannot be "worked around," because they focus on function rather than form. In either case, it should be kept in mind that a patent reflects the fact that someone, the patent holder, has been identified as the *owner* of this intellectual property.

A computer-based version of an index to patents, *Classification and Search Support Information System* (CASSIS), can be found at most libraries. Patents are listed by individual class and subclass numbers that are detailed in a rather complex classification index. The USPTO maintains its own Web site (and search engine) that presents data about granted patents, and it publishes a weekly edition of the *Official Gazette* that lists, in numerical order by class number, all patents granted in the previous week.

6.2.1.3   Group Activities for the Design Team

We have already identified brainstorming as an activity done by a relatively small number of people, working in a group setting, while clarifying the clients' goals. In Section 3.5.1 we pointed out that new ideas, some related, some not, might be generated, and we recommended that these new ideas be recorded or listed (although we also cautioned against evaluating them at that time). We also suggested that design team members show significant respect for the ideas and offerings of their teammates. In the context of generating design alternatives, we now extend those behavioral themes for design teams in action as we emphasize some "rules of the game" for three different "games" that teams can play in order to generate design ideas.

Successful design calls for two different kinds of thought, divergent thinking and convergent thinking. When we are getting together to expand the design space and generate design alternatives, we want to "think outside of the box" or "stretch the envelope." In other words, here we want to do *divergent thinking* because we want to remove limits or barriers, hoping instead to be expansive while we are trying to increase our store of design ideas and choices. On the other hand, after we have opened up the design space sufficiently, we then want to narrow that space to focus on the best alternative(s). Here it becomes useful to do what is called *convergent thinking*, that is, to do our thinking and problem solving rather narrowly, so as to converge to a solution within known boundaries or limits.

(As an aside, we note that design, a goal-oriented activity, combines divergent and convergent thinking in some interesting and complex ways. As an overall process, for example, it seeks to converge upon the best possible solution, however that is defined and measured. At the same time, the various activities that lead us to finally converge on a solution require a great deal of divergent thinking. At the earliest stages, for example, when we seek to understand the problem and identify other stakeholders who have an interest in the solution, we clearly want to open up the problem space. Similarly, we later engage in divergent thinking when we expand our functions to cover the most general case. At the end of each of these sets of "divergent" activities, we try to converge to an adequate representation of our understanding. Thus, setting and documenting priorities among objectives is an essentially convergent process, as is selecting appropriate metrics for our objectives. This interplay of modes of thinking and of acting is part of what gives design its intellectual richness.)

The three games or activities we now describe are intuitive in nature because they focus on ways to encourage divergent thinking. Rather than presenting a strict logical (or axiomatic, or algorithmic) approach, we strive here to encourage "free thinking" in order to enhance our collective creativity. Please do note, though, that we do not mean to encourage illogical thinking, or thinking that violates physical principles or the axioms of logic. We are, instead, recognizing that people working in groups often can interact more spontaneously, in a free-spirited way that calls forth associations from group members in ways that we cannot anticipate or logically force. These activities may also be characterized as *progressive* in nature because there are iterative cycles within each that results in the progressive emergence and refinement of new design ideas.

The first activity or group design game or method is called the *6–3–5 method*. The name derives from the idea of having six team members seat themselves around a table in order to participate in this idea generation game. Each participant writes an initial list of three design ideas that are cast as brief expressions in key words and phrases. Each individual list is then circulated past all of the remaining group members for *written* comment and annotation, as we would allow no verbal communication or cross-talk, in a sequence of five rotations. Thus, each list travels a complete circuit around the table, and each member of the group is stimulated in turn by the increasingly annotated lists of the other team members. When all of the participants have commented on each of the lists, the team can then use a blackboard or some other visualization tool to list, discuss, evaluate, and record all of the design ideas that have resulted from this serial, group enhancement of all of the team members' individual ideas.

Note that we could easily generalize this method to the "m–3–(m-1)" method by starting with m team members and using m-1 rotations to complete a cycle. However, the logistics of dealing with ever-lengthening lists written on increasingly crowded sheets of paper, and tables seating more than six people, suggest that six may be a "natural" upper limit for this activity, Remember, too, that we would generally have six or fewer people—ideally no more than four—on a project team, especially in an academic environment.

The *C-sketch method* is similar to the 6–3–5 method except that its starting point is the initial presentation of a single design concept as a sketch by each of the team members. The sketches are then circulated through the team in the same fashion as the lists of ideas in the 6–3–5 method, with all of the annotations or proposed design modifications being written or sketched on the initial concept sketches. Again, the only permissible channel of communication during the process is through pencil on paper, and the discussions that follow the end of a complete cycle of sketching and modifying follow much the same form as described for the 6–3–5 method. Research does suggest that the C-sketch method becomes very unwieldy with even five team members because of the crowding of annotations and modifications on a given sketch. Finally, it is worth noting that the C-sketch method is very appealing in an area such as mechanical design. We have already noted that there is strongly suggestive evidence that sketching is a natural form of thinking in mechanical device design, and research has also shown that drawings and diagrams facilitate the grouping of relevant information that is added in the form of marginal notes, and they help people to better visualize whatever objects are being discussed.

Finally, the *gallery method* is another approach to getting team reactions to drawings and sketches, although the sketching and communication cycles are handled differently. In the gallery method, group members first develop their individual, initial ideas within some allotted time, after which all of the resulting sketches are posted, say on a corkboard or a conference room whiteboard. This set of sketches forms the backdrop for an open, group discussion of all of the posted ideas. Here questions can be asked, critiques offered, and suggestions made. Then each participant returns to her or his drawing and suitably modifies or revises it, again within a specified period of time, with the goal of producing a second generation idea. Obviously, the gallery method is also an iterative, progressive approach, and there is no way to predict just how many cycles of individual idea generation and group discussion should be conducted. Our only recourse would be to apply the common sense dictum of the law of diminishing returns, that is, proceed until a concensus develops within the group that one more cycle would not produce much (or any) new information, at which point it is time to quit because it wouldn't make sense to expend any further effort on this particular activity.

6.2.1.4   Synectics and Other Ways to Think Divergently

Having described some ways in which a team can conduct itself, it is useful for us to describe some of the kinds of things that team members can think about. These frameworks for thinking, often called *synectics*, can also be viewed as recipes for divergent and expansive thinking that are intended to enhance and advance the collective creativity of a design team. The dictionary defines synectics as "a theory or system of problem-stating and problem-solution based on creative thinking that involves free use of metaphor and analogy in informal interchange within a carefully selected small group of individuals of diverse personality and areas of specialization." Of course, there is no guarantee that all design teams will have the right personal ingredients. And while all of the ideas described can (and should) go on in the minds of each team member, it is their application in a group setting that generally reaps the most benefits.

A *metaphor* is a figure of speech, that is, it is a style where we use attributes of one object or process to give depth or color to the description of a second object or process. For example, to describe the study of engineering as drinking from a fire hose is to suggest that engineering students are expected to absorb a great deal of knowledge rapidly and under great pressure. We use metaphors when we want to point out *analogies* between two different situations, that is, when we want to suggest that there are parallels or similarities in the two sets of circumstances. Analogies can be very powerful tools in engineering design. One of the most frequently cited is the case of Velcro fasteners, for which the analogy drawn was to those pesky little plant burrs that seem to stick to everything they're blown on to.

The Velcro fastener is the result of a *direct analogy*, wherein the inventor made a direct connection between the individual elements of the plant burr and the connecting fibres of the fastener. We also can use *symbolic analogies*, as when we plant ideas or talk about objectives trees. In these cases we are clearly drawing connections through some underlying symbolism. We might also apply *personal analogies* by imagining what it would feel like to be the object (or part of it) that we're trying to design. For example, how might it feel to be a tin beverage container with a pull-tab? We could also stray into the realm of *fantasy analogies* by imagining something that is, literally, fantastic or beyond belief. Of course, looking at our world at the end of the second millennium, just how much more fantastic can it get than space travel, instantaneous and reliable personal communication across the globe, and seeing clearly into the human body with CT scans and magnetic resonance imaging (MRI).

Fantasy analogies suggest another approach to "thinking out of the box." We are not very far past the time when some of the technologies that we now take for granted were thought to be outrageous ideas that were simply beyond belief. For example, when Jules Verne's classic *20,000 Leagues Under the Sea* was published in 1871, the idea of ships that could "sail" deep in the ocean was viewed as an out-rageous fantasy. Now, of course, the notions of both submarines and of seeing unfamiliar yet exciting lifeforms under water are part of our everyday experience. (And Verne's submarines are not unique: remember the attacking spaceships in H. G. Wells' *War of the Worlds*.) Indeed, it has often been said that science fiction writers are the world's most prescient futurists. Whether that's true or not, we cannot escape the idea that design teams might imagine the most outrageous solutions to a design problem and then seek ways that such solutions could be made useful. For example, airplanes that are invisible to radar were once consid-ered far fetched. The arterial stents used in angioplastic surgery (see Figure 6.1) are still another example of devices once thought to be impossible. Who would have believed that we could erect an engineering structure within the narrow (3–5 mm in diameter) confines of a human artery?

The stent suggests still another aspect of analogical thinking, namely, that it is often quite useful to look for *similar solutions*. The stent clearly is similar in both intent and function to the scaffolding that is erected to support walls in mines and tunnels as they are being built. Thus, the stent and the scaffold are *like ideas*.

We could invert this idea by looking for *contrasting solutions* in which the conditions are so different, so contrasting, that a transfer of solutions would seem totally implausible. In other words, here we would be looking for *opposite ideas*. Fairly obvious contrasts would be between strong and weak, light and dark, hot and cold, high and low, and so on. One example of using an opposite idea occurs in guitar design. Most guitars have their tuning pegs arrayed at the end of the neck. In order to make a portable guitar, one clever designer chose to put the tuning pegs at the other end of the strings, at the bottom of the body, in order to save space and thus accentuate the guitar's portability.

Finally, in addition to looking for similar and contrasting solutions, we could recognize a third category. *Contiguous solutions* are developed by thinking of *adjoining* (or *adjacent*) *ideas* in which we take advantage of natural connections

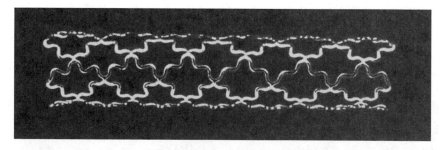

Figure 6.1 This is a Palmaz-Schatz™ balloon expandable coronary *stent,* that is, a device used to maintain arterial shape and size so as to allow uninhibited and natu-ral blood flow. Note how this structure resembles the kind of scaffolding often seen around building renovation and construction projects. (Courtesy of Cordis, a Johnson & Johnson Company.)

between ideas or concepts or artifacts. For example, a chair might prompt us to think of a table, as a tire might prompt us to think of a car, and so on. One way to distinguish contiguous solutions from similar solutions is to think about the adjacency of a bolt to a nut, as contrasted with the fact that bolts and rivets serve identical fastening functions.

Again, it is worth keeping in mind that group activities oriented toward generating ideas and expanding the design space work best if the environments in which they are conducted are open, positive, and flowing. The important environmental issues are less about the chairs and the whiteboards than they are about intellectual openness and emotional support. Every participant should feel free and be free to take risks, to offer ideas that might in other circumstances seem far-fetched or foolish, and to be as welcoming to "wild" ideas of others in the group as they are sensitive about the reception given their own ideas.

6.2.2   Limiting the Design Space to a Useful Size

On the more general subject of guiding the search for design solutions, there are some pragmatic issues relating to how an artifact will be both used and made that ought to be used as guideposts while developing a search space. However, it must be said that these guideposts may not always expand the search space. In fact, they may narrow the search space. These guideposts are, in particular, to analyze design alternatives as "functions" of:

- user needs
- available technologies
- external constraints

For example, the design of vehicles for a campus transportation system could be substantially influenced by the *user needs* of potential users. Candidate vehicles could include low-end, simple bikes; nifty, high-tech bikes; recumbent bikes; tricycles; or even rickshaws. Among the user needs that might affect the consideration and design of these vehicle types would be the availability of parking facilities at lecture and residence halls, the need to carry packages, a need for access for the handicapped, and so on. User needs could have a major impact on the kinds of features called for in a campus vehicle, and thus on the very vehicle type. Just as clearly, the *availability of different technologies* could stimulate the process of generating alternatives. Just think of the different materials of which a bike or bike-like device could be made. The choice of materials would certainly affect a bike's appearance, its manufacture, and its price. And lastly, *external constraints* such as the team's competence in certain design fields (e.g., they may be more comfortable designing inexpensive tricycles than high-tech performance bikes) or the availability of manufacturing capabilities (e.g., avoid designing a bike made of composite materials if the only manufacturing facility available is one that can form and connect metals).

Similarly, there are some practical considerations to keep in mind while building a design space so that its size and scope stay both manageable and useful. These considerations are, in fact, largely issues of *common sense* wherein we should again follow Keynes' aphorism: "Nothing is required, and nothing will avail, but a little, a very little clear thinking." In particular, lest the lists of issues and of candidate designs become too large or too silly, we can:

- *invoke and apply constraints*, in much the same way we did just above while trying to assess the influence and importance of user needs
- *freeze the number of attributes* under consideration, often just to avoid getting into levels of detail that are unlikely to seriously affect the design at this point (e.g., the color of a bike or car is hardly worth noting as a design objective in the early stages of design)
- *impose some order on the list*, perhaps by harking back to our weighted objectives trees or to other data gathered in the problem definition phase that might suggest that some functions or features are more important
- *"get real!"* or, in other words, watch out when silly, unfeasible ideas are posed, although, again, such common sense must also be applied at a time and in a style consistent with our earlier admonitions about conducive and supportive environments for brainstorming

We will have more to say about managing the process of generating and selecting design alternatives in Section 6.7.

6.3 MORPHOLOGICAL CHARTS: ORGANIZING FUNCTIONS AND MEANS TO GENERATE IDEAS THAT REALLY WORK

Morphological charts are another in our series of charts, matrices, and tables that represent planar spaces in which we can visualize elements of our design space. In fact, *morph charts*, as they are often (and affectionately) called, show a view of the design space that allows us not only to identify potential designs, but also to get a sense of the size of the design space.

We begin by constructing a list, much as we did for building objectives trees, but here we start either with features we want the design to have or functions that we want it to perform. This list of functions or features should be of a reasonable, manageable size, and at the same level of detail (or of abstraction). That is, we should analyze functions or features that are at the same level within objectives or function–means trees as this helps ensure internal consistency. For each function or feature identified, we make a sublist of all of the different means of implementing that feature or function. Thus, for example, a list of features and functions for the beverage container problem that is consistent with the subobjectives of the objective *Promote Sales* (see Figure 3.2) might look like:

- Contain Beverage
- Material for Beverage Container
- Provide Access to Juice
- Display Product Information
- Sequence Manufacture of Juice and Container

If we now build sublists of means for achieving each of these ends or functions and attach them to the corresponding features and attributes, we would find the following table:

Contain Beverage:	Can, Bottle, Bag, Box
Material for Beverage Container:	Aluminum, Plastic, Glass, Waxed Carboard, Lined Cardboard, Mylar Films
Provide Access to Juice:	Pull-Tab, Inserted Straw, Twist-Top, Tear Corner, Unfold Container, Zipper
Display Product Information:	Shape of Container, Labels, Color of Material
Sequence Manufacture of Juice and Container:	Concurrent, Serial

The information in this table provides what we need to construct a morph chart. In fact, we show the resulting *morphological chart* in Figure 6.2, wherein we see the information just given is displayed in a form that is both visually appealing and usefully organized. The features and functions that the device must serve are listed on the vertical axis, while for each of these attributes two or more means are identified and listed in cells in that row. A conceptual design or scheme can be constructed by linking one means, any means, for each of the (five) identified functions, subject only to interface constraints that may prevent a particular combination. For example, one design could consist of a plastic bottle with a twist-top, its color chosen to correspond to a particular beverage, and made (and stored) in advance of the expected delivery of the beverage. We are clearly assembling a design in the classic "Chinese menu" style, choosing one means from each of rows $A, B, C \ldots$ to combine into a design scheme. Similarly,

Figure 6.2 A morphological chart for the beverage container design problem. The functions that the device must serve are listed on the vertical axis, while for each of them two or more means are identified. Subject only to interface constraints that may prevent a particular combination, a conceptual design or scheme can be constructed by linking one means, any means, for each of the five identified functions, thus assembling a design in the classic "Chinese menu" style.

MEANS FEATURE/ FUNCTION	1	2	3	4	5	6
Contain Beverage	Can	Bottle	Bag	Box	••••	••••
Material for Drink Container	Aluminum	Plastic	Glass	Waxed Cardboard	Lined Cardboard	Mylar Films
Mechanism to Provide Access to Juice	Pull Tab	Inserted Straw	Twist Top	Tear Corner	Unfold Container	••••
Display of Product Information	Shape of Container	Labels	Color of Material	••••	••••	••••
Sequence Manufacture of Juice, Container	Concurrent	Serial	••••	••••	••••	••••

we can see the morph chart as being somewhat analogous to a spreadsheet and similarly enabling certains kinds of "calculations."

How many potential solutions are thus identified in a morph chart? Or, in other words, just how big is our design space getting to be? A simplistic—and wrong—answer would be to say that the design space, expressed in terms of the number of candidate designs, is simply the product of the maximum number of means times the number of functions and features. This answer is wrong, however, because it does not account for the *combinatorics* that result from combining any single means in a given row with each of the remaining means in all of the other rows. Thus, for the beverage container morph chart of Figure 6.2, the number of candidate designs could be as large as $4 \times 6 \times 6 \times 3 \times 2 = 864$.

However, having done this calculation and seen how large a design space can become, it is important to recognize that not all of these 864 connections are, in fact, feasible or candidate designs. For example, while it may seem laughingly obvious, we are unlikely to design a glass bag with a zipper! We have to learn, therefore, that while building a morphological chart provides a way to expand our design space and identify alternatives, it also provides an opportunity to *prune the design space by identifying and excluding incompatible alternatives*. We cannot assume that all of the connections made by following out the combinatorial arithmetic are valid connections. There are clearly other combinations of container designs that cannot work (e.g., glass cans with pull-tabs or inserted straws) and therefore must be excluded from the design space. To exclude such alternatives we can apply design constraints, physical principles, and plain common sense. We should also remember that technologies and, consequently, available means, do change over time. For example, the waxed paper container has evolved from being one that really only supported the containment function, to a modern one that incorporates a twist-cap at its top, which means that this container supports both containment and intermittent mixing (after it has been opened) of the contents. (And that's why we have these tops on orange juice cartons, but not on milk cartons.)

We noted above that it was important to list our features and functions at the same level of detail when building a morph chart. After all, we don't want to compare apples and oranges. Thus, for the beverage container, we would not include means of *Resisting Temperature* and means for *Resisting Forces and Shocks* within the morph chart of Figure 6.2 because they are more detailed functions that derive from subgoals that are much further down in the objectives tree of Figure 3.2. Similarly, when doing a complex design task, such as designing a building, we don't want to worry about the means for identifying exits or for opening doors at the same time that we are trying to develop different concepts for moving between the ground floor and the shopping mezzanine, which might include elevators, escalators, and stairways.

As a further illustration, we show in Figure 6.3 a morph chart constructed in a freshman design project done at Harvey Mudd, this one being concerned

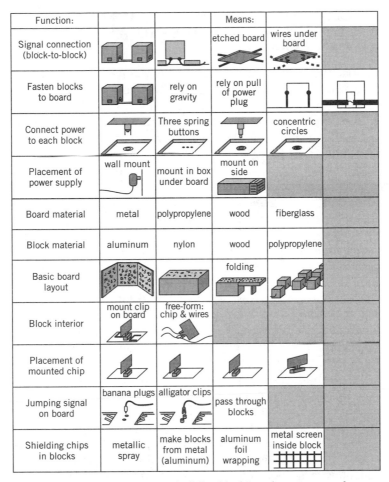

Function:			Means:		
Signal connection (block-to-block)			etched board	wires under board	
Fasten blocks to board		rely on gravity	rely on pull of power plug		
Connect power to each block		Three spring buttons		concentric circles	
Placement of power supply	wall mount	mount in box under board	mount on side		
Board material	metal	polypropylene	wood	fiberglass	
Block material	aluminum	nylon	wood	polypropylene	
Basic board layout			folding		
Block interior	mount clip on board	free-form: chip & wires			
Placement of mounted chip					
Jumping signal on board	banana plugs	alligator clips	pass through blocks		
Shielding chips in blocks	metallic spray	make blocks from metal (aluminum)	aluminum foil wrapping	metal screen inside block	

Figure 6.3 A morph chart for a "building block" analog computer that was done in Harvey Mudd's E4 design course (Hartmann, Hulse et al. 1993).

with the design of a "building block" analog computer. This morphological chart is particularly appealing in its use of graphics and icons to illustrate many of the means that could be applied to achieve functions in this design problem. This style of morph chart may also be particularly useful in the kind of C-sketch discussions described in Section 6.2.1.3.

Along similar lines, we note that morph charts can also be used to expand the design space for large, complex systems by listing principal subsystems as our starting column and then identifying various means of implementing each of the subsystems. For example, if we were designing some sort of vehicle, one subsystem would be its *Source of Power*, for which the corresponding means could be *Gasoline, Diesel, Battery, Steam,* and *LNG*. Each of these sources of power is itself a subsystem that needs further detailed design, but the array of power subsystems expands the range of our design choices.

6.4   SELECTING THE BEST ALTERNATIVES: CONNECTING ALTERNATIVES TO WEIGHTED OBJECTIVES AND METRICS

If we have done the job of generating design alternatives properly, we almost certainly have several design concepts or alternatives from which to choose. We may have used a morph chart to identify a (potentially large) number of design combinations, or we may have generated design alternatives using a less-structured approach. No matter how we've done that, we now have to choose from the identified options and select one, perhaps two, concepts or designs for further development and testing. We are forced to choose only one or two simply because resources of time, money, or personnel are never infinite. Thus, we now focus on ways that we can "pick a winner" and choose the design best suited for further elaboration, testing, and evaluation. The methods we describe here all turn on the use of the weighted objectives that we described in Chapter 3. This is clearly a sensible connection because our designs should meet the objectives of all of the various stakeholders, including clients, users, the public at large, and so on.

There are many approaches we can use to assess our design alternatives in terms of our weighted objectives, ranging from formal and rigorous methods to simply picking the one(s) we "like" the best. We will discuss three methods that explicitly link design alternatives to objectives, which enables us to rank alternative designs in terms of those objectives. We might note that the methods we suggest for selecting a "best" design do not have the mathematical or algorithmic certainty that goes along with finding maxima and minima in calculus. Rather, here we are striving as best we can to bring as much order as possible to judgments and assessments that are, at their root, subjective. Just as professors give letter and number grades to encapsulate judgments about how well students have mastered concepts, ideas, or methods, designers are trying to integrate the best judgments of the stakeholders and, especially, other design team members, in ways that allow these judgments to be exploited in a sensible and orderly manner. On the other hand, this lack of mathematical rigor should not be taken as a license to accept the results calculated without thinking them through. That is, we must use our common sense when we look at the results of the methods we now describe.

There are, as noted, three methods that we can use to select one from among a set of alternative designs or concepts: the numerical evaluation matrix, the weighted checkmarks chart, and the best of class chart. No matter which one of the charts we choose to apply, our first step should always be to examine and evaluate each alternative in terms of all of the constraints that apply. Recall that alternative designs must be rejected if the constraints are not met. As we outline our three selection methods, we will assume that all applicable constraints are being applied and that the design space is being narrowed accordingly.

We show numerical design *evaluation matrices* for the BJIC and GRAFT companies in Figures 6.4 and 6.5. These charts show objectives and their relative weights in the left-hand columns, while the scores assigned to every objective as a function of particular design are shown in design-specific columns on the right.

Figure 6.4 A numerical evaluation matrix for the beverage container design problem. This chart reflects BJIC's values in terms of the weights assigned to each objective, which are the same as those in the weighted objectives tree of Figure 3.6. They also correspond to the results given in the pairwise comparison chart of Figure 3.4 (b).

DESIGN CONSTRAINTS/ OBJECTIVES	Weight (%)	Glass bottle, with twist-off cap	Aluminum can, with pull-tab	Polyethylene bottle, with twist-off cap	Mylar bag, with straw
C: No sharp edges		✗	✗		
C: No toxin release					
C: Preserves quality					
O: Environmentally benign	33			0.9 ǀ 33% 29.7%	0.1 ǀ 33% 3.3%
O: Easy to distribute	09			0.5 ǀ 9% 4.5%	0.6 ǀ 9% 5.4%
O: Preserves taste	22			0.9 ǀ 22% 19.8%	1.0 ǀ 22% 22%
O: Appeals to parents	18			0.8 ǀ 18% 14.4%	0.5 ǀ 18% 9.0%
O: Permits market- ing flexibility	04			0.5 ǀ 4% 2.0%	0.5 ǀ 4% 2.0%
O: Generates brand identity	13			0.2 ǀ 13% 2.6%	1.0 ǀ 13% 13%
TOTALS	99			73.0%	54.7%

The constraints for this beverage container problem are shown at the top of the charts, and as a result of applying them, we would judge glass bottles and aluminum containers to be unacceptable because of potential sharp edges. This reduces the number of alternatives to two: the Mylar bag and the polyethylene bottle. We now score these two alternatives using the metrics discussed in Chapter 5 and the weighted objectives developed in Chapter 3.

Thus, for example, a polyethylene bottle scores 0.9 for *environmentally benign,* while the Mylar bag earns a score of 0.1 for this metric. (Note, too, that all of the metrics results have been normalized to a 0–1 range.) Since the BJIC Company values *environmentally benign* much more (i.e., 33%) than the GRAFT Company (i.e., 4%), the scores earned for each candidate design for this metric are significantly less for BJIC than for GRAFT. When we look at the cumulative results for all of the objectives for the two remaining viable products, we see that BJIC's values effectively rate the polyethylene bottle significantly ahead, while GRAFT's values dictate a choice of the Mylar bag by a similar margin.

Figure 6.5 A numerical evaluation matrix for the beverage container design problem. This chart reflects GRAFT's values in terms of the weights assigned to each objective, as shown in the pairwise comparison chart of Figure 3.4 (a). However, note that the scores found for each metric in the chart are the same as those used for the BJIC design and shown in Figure 6.4. Is that as it should be? If so, why?

DESIGN CONSTRAINTS/ OBJECTIVES	Weight (%)	Glass bottle, with twist-off cap	Aluminum can, with pull-tab	Polyethylene bottle, with twist-off cap	Mylar bag, with straw
C: No sharp edges		✗	✗		
C: No toxin release					
C: Preserves quality					
O: Environmentally benign	04			0.9 ǀ 4% 3.6%	0.1 ǀ 4% 0.4%
O: Easy to distribute	22			0.5 ǀ 22% 11.0%	0.6 ǀ 22% 13.2%
O: Preserves taste	09			0.9 ǀ 9% 8.1%	1.0 ǀ 9% 9%
O: Appeals to parents	13			0.8 ǀ 13% 10.4%	0.5 ǀ 13% 6.5%
O: Permits market- ing flexibility	18			0.5 ǀ 18% 9.0%	0.5 ǀ 18% 9.0%
O: Generates brand identity	33			0.2 ǀ 33% 6.6%	1.0 ǀ 33% 33%
TOTALS	99			43.7%	74.7%

Beyond these calculated results about these two hypothetical candidate designs, the most important feature in Figures 6.4 and 6.5 is that each chart has the same value for the metrics applied to the design alternatives. Recall from our discussion in Section 5.3 that metrics are measurable indicators of how well specific objectives are met. Thus, if our metrics had different values for different design alternatives, we would have to assume that the testing process is defective. Here the design team has clearly selected different alternatives as a proper reflection of the fact that their different clients (i.e., BJIC and GRAFT) have weighted their objectives differently, perhaps because they have different corporate values. There was no difference in the design selection process in either the metrics or the testing procedures.

However, it is worth noting that this might not be the case if the companies were independently doing their designs and, consequently, rating each product on its different dimensions. That is, it is not at all hard to imagine that some companies might find a Mylar bag significantly more expensive to produce and distribute than they would a polyethylene bottle. In such a case, the metric for low cost of production and distribution might be lowered from 0.6 to 0.1, for example, from which a rather different outcome would emerge.

It is also worth noting that applications of the weighted objectives chart can produce a lot of detailed, numerical results, especially if the number of objectives is high and weighted results are carried out for a large number of concepts or alternatives. In some cases, such as large public infrastructure projects, such detail is not only appropriate, it is almost certainly necessary to publicly document the design process that was followed. Public highway projects, for example, often turn out to be quite contentious, as a result of which public agencies often try to pay attention to a large number of stakeholders, who have many and differing objectives, and for which the range of potential solutions may also be large. We will see an instance of this in Chapter 10 when we describe the Charlestown, Massachusetts, City Square project. However, this level of detail may not be needed in other design projects, so that the next two charts may suffice.

In the *weighted checkmark method*, we first simply rank the objectives as high, medium, or low in priority. Objectives with high priority are given three checks, those with medium priority are given two checks, while objectives with low priority are given only one check, as shown in Figure 6.6. Similarly, metrics are taken as 1 if rated greater than 0.5, and as 0 if their rating is less than 0.5. If a

Figure 6.6 A weighted benchmark chart for the beverage container design problem. This chart qualitatively reflects BJIC's values in terms of the weights assigned to each objective, so it is a qualitative version of the evaluation matrix of Figure 3.4.

DESIGN CONSTRAINTS/ OBJECTIVES	Weight (✓)	Glass bottle, with twist-off cap	Aluminum can, with pull-tab	Polyethylene bottle, with twist-off cap	Mylar bag, with straw
C: No sharp edges		✗	✗		
C: No toxin release					
C: Preserves quality					
O: Environmentally benign	✓✓✓			1❙✓✓✓ ✓✓✓	0❙✓✓✓ ••••
O: Easy to distribute	✓			1❙✓ ✓	1❙✓ ✓
O: Preserves taste	✓✓			1❙✓✓ ✓✓	1❙✓✓ ✓✓
O: Appeals to parents	✓✓			1❙✓✓ ✓✓	1❙✓✓ ✓✓
O: Permits marketing flexibility	✓			1❙✓ ✓	1❙✓ ✓
O: Generates brand identity	✓✓			0❙✓✓ ••••	1❙✓✓ ✓✓
TOTALS				9✓	8✓

design alternative meets an objective in a "satisfactory" way, it is then marked with one or more checks, as shown in Figure 6.6. Finally, the number of checks are summed over all of the alternatives, all of the constraints having already been applied. This method is very easy to use, makes the setting of priorities rather simple, and is readily understood by clients and by other parties. On the other hand, the weighted checkmark approach lacks detailed definition and it sets up all of the metrics as binary variables, that is, they are either checked (satisfactory) or they are not. This makes it much easier to succumb to temptation and "cook the results" in order to achieve a desired outcome.

Our last method for evaluating and ranking alternatives is the *best of class chart*. For each objective, we assign increasing scores to each design alternative that range from 1 for the alternative that meets that objective best, 2 to second best, and so on, until the alternative that met the objective worst is given a score equal to the number of alternatives being considered. If, for example, there are 5 alternatives, then the best at meeting a particular objective would receive a 1, the second best a 2, etc. Ties can be allowed (e.g., there are two alternatives that are considered equally "best" and so are tied for first). Ties would be handled by splitting the available rankings (e.g., the two "firsts" would each get a score of $(1+2)/2 = 1.5$). These scores are then weighted in accord with the objectives' weights already found, just as we have done with the weighted objectives charts, the balance of the calculation would proceed as in Figures 6.4 and 6.5, and the *lowest* summed score would be considered to be the best alternative design under this scheme.

The best of class approach also has its advantages and its disadvantages. One advantage is that it allows us to rank alternatives with respect to a metric, rather than simply treat as a binary, "yes or no" decision, as we did just above. It, too, is relatively simple to implement and to understand, and it can be done by individual team members or by a team as a group in order to make explicit any differences in rankings or approaches. The disadvantages of this approach are that it encourages evaluation based on opinion rather than testing or on the actual metrics, and it may lead to a moral hazard similar to that attached to weighted checkmarks, that is, the temptation to fudge the results or cook the books.

No matter which of these three methods is used, there is no excuse for accepting the results blindly and uncritically. First of all, common sense must be applied when we are evaluating results. If two alternatives are relatively close, then it is almost certainly unwise to treat them as anything other than tied, unless further thinking uncovers unevaluated strengths or weaknesses or untested metrics. Second, if the evaluation results are unexpected, then we need to ask whether our expectations were simply wrong, whether the method was consistently applied, or whether our weights are not really appropriate to the problem. Third, if the results are as expected, we should ask whether or not our results truly represent a fair application of the evaluation process, or have we just reinforced some preconceived ideas or biases. Finally, if some alternatives have been rejected because they violated constraints, it would be wise to ask whether those constraints are truly binding.

6.5   PROTOTYPES, MODELS, AND PROOFS OF CONCEPT

In this section we discuss three-dimensional, physical realizations of concepts for designed artifacts. In other words, we want to talk about things that are being made to at least strongly resemble the object being designed, if not actually mimic what is supposed to be "the real thing." There are several versions of physical things that could be made, including prototypes, models, and proofs of concept, and they are often made by the designer.

Prototypes are, according to the dictionary, "original models on which something is patterned." They are also seen as the "first full-scale and usually functional forms of a new type or design of a construction (as an airplane)." In this context, clearly, prototypes are working models of designed artifacts. They are tested in the operating environments in which they're expected to function as final products. It is interesting that aircraft companies routinely build prototypes, while rarely, if ever, does anyone build a prototype of a building.

A *model* is "a miniature representation of something," or a "pattern of something to be made," or "an example for imitation or emulation." We use models to *represent* some devices or processes. They may be paper models or computer models or physical models. We use them to illustrate certain behaviors or phenomena as we attempt to verify the validity of an underlying (predictive) theory. Further, models are often smaller and made of different materials than are the original artifacts they are intended to represent, and they are typically tested in a laboratory or in some other controlled environment in order to validate their expected behavior.

Our definitions and initial discussions of prototypes and models sound similar enough that it is worth asking: Are prototypes and models the same thing? The answer is, "not exactly." As we explain below, the distinctions between prototypes and models may have more to do with intent behind their making and the environments in which they are tested than with any clearly definable dictionary-type differences. Prototypes are intended to demonstrate that a product will function as designed and so they are tested in their actual operating environments or in similar, uncontrolled environments that are as close to their relevant "real worlds" as possible. Models are intentionally tested in controlled environments that allow the model builder (and the designer, as they may not be the same person) to understand the particular behavior or phenomenon that is being modeled. An airplane prototype is made of the same materials and has the same size, shape, and configuration as those intended to fly in that series (i.e., Boeing 747s or Airbus 310s). Clearly, a model airplane would likely be (much) smaller; it might be "flown" in a wind tunnel or for sheer enjoyment, but it is not a prototype.

Engineers often build models of buildings, for example, for the wind-tunnel testing of proposed skyscrapers, but these models are not prototypes. Rather, the building models used in a wind-tunnel simulation of a cityscape with a new high rise are essentially toy building blocks that are meant to imitate the skyline. They are not buildings that work in the sense of aircraft prototypes that actually fly. So, why is it that aeronautical engineers build prototype airplanes, while civil engineers do not build prototype buildings? And what do they do in the other disciplines?

The answer is that it depends. The decision to build a prototype depends on a number of issues, including: the size and kind of the design space, the costs of building a prototype, the ease of building that prototype, the role that a full-size prototype can play in ensuring the widespread acceptance of a new design, and the number of copies of the final artifact that are expected to be made or built. Aircraft and buildings are interesting examples here, because they provide both ample commonalities and sharp differences. The design spaces of commercial aircraft and of high-rise buildings are large and complex. There are, literally, millions of parts in each, so there are many, many design choices to be made along the way. The costs of building both airplanes and tall buildings are also quite high. In addition, at this point in our cultural advancement, we have ample experience with both aeronautical and structural technologies, so that we generally have a pretty good idea of what we're about in these two domains. So, again, why prototype aircraft and not prototype buildings? In fact, don't the complexity and expense of building even a prototype aircraft argue directly against the idea of building such prototypes?

Notwithstanding all of our past experience with successful aircraft, we build prototypes of airplanes because, in large part, the chances of a catastrophic failure of a "paper design" are still unacceptably high, especially for the highly regulated and very competitive commercial airline industry that is the customer for new civilian aircraft. That is, we are simply not willing to pay the price of having a brand new airplane take off for the first time with a full load of passengers, only to watch hundreds of lives being lost—as well as the concomitant loss of investment and of confidence in future variants of that particular plane. Thus, it is in part an ethical issue, because we do bear responsibilities for our technical decisions when they impinge upon our fellow humans. It is also in part an economic issue, because the cost of the prototype is economically justifiable when weighed against potential losses. Further, we also build prototypes of airplanes because those particular planes are not simply thrown away as "losses" after testing; they are retained and used as the first in the series of the many full-size designs that are the rest of the fleet of that kind of airplane.

To be sure, there are catastrophic failures of buildings, both during and after construction. However, they are so comparatively rare in number that there is little perceptible value in requiring that prototype buildings be built before occupancy. Building catastrophes are rare in part because high rises can be tested, inspected, and experienced gradually, as they are being built, floor by floor. The continuous inspection that takes place during the construction of a building, of each step from the foundation on up, certainly has its counterpart in numerous inspections and certifications that accompany the manufacture and assembly of a commercial airliner. But, finally, the maiden flight of an airplane is a binary issue, that is, the plane either flies or it doesn't, and any failure is not likely to be a graceful degradation!

One interesting aspect of comparing the design and testing of airplanes with that of buildings has to do with the number of copies being made. We have already noted that prototype aircraft are not simply discarded after their initial test flights, but they are flown and used. It is also the case that airframe manufacturers are in business to build and sell as many copies of their prototype aircraft as they can, so engineering economics plays a role in the decision to build a prototype. The economics are complicated because the manufacturing costs of building the first plane in a series are very high. There are numerous technical decisions to be made about the kinds of tooling and the numbers of machines needed to make an airplane, and there are economic tradeoffs to be evaluated between the investment needed to fund the manufacturing process and the anticipated revenue from the sale of the aircraft. We will address some of the manufacturing cost issues in Chapter 8. (And as an interesting little aside, we might say that residential housing developments could be viewed as having prototype buildings since they all have "model homes," although it's a bit hard to imagine the marketing success of "a fully furnished prototype home.")

Another lesson we can learn from thinking about buildings and airplanes is that there is no obvious correlation between the size and cost of prototyping— or the decision to build a prototype—and the size and type of the design space. And while it might seem that the decision to build a prototype might be strongly influenced by the relative ease of building it, the aircraft case shows that there are times when even costly, complicated prototypes must be built. On the other hand, and in general terms, if it is cheap and easy to do, then it would seem like a pretty good idea to build a prototype. Certainly there are instances where prototypes are commonplace, for example, in the software business. Long before a new program is shrinkwrapped and shipped, it is alpha- and beta-tested as early versions are prototyped, tested, evaluated, and, hopefully, fixed. In fact, we might see the evolution of version numbers to forms such as XXX 5.25 as indicators that even those who bought XXX 5.17 as a finished product may have actually been buying an older prototype.

If there is a single lesson about prototypes, beyond that it is generally good to build them, it is that plans for doing so should be made in both the project WBS and in the budget. More often than not a prototype is required, although there may be instances in which neither resources nor time are available. In weapons development contracts, for example, the U.S. Department of Defense virtually always requires that a design concept be demonstrated so that its performance can be evaluated before large-scale procurement is ordered. At the same time, it is interesting to report that the large aircraft companies are discussing the idea that advances in computer-aided design and analysis, as well as in our understanding of aircraft and their operating environments, may have made it possible to dispense with prototype development and replace it with sophisticated simulation. The potential cost savings are clearly enormous, and very tempting, but the implications are also vast and uncertain. For our purposes, however, it still remains true that prototyping is generally preferred.

Sometimes we build prototypes of parts of large, complex systems and use them as models to check how well those parts behave or function. For example, structural engineers have been known to build full-size connections, say, at a point where several columns and beams intersect in a geometrically complicated way, and test them in the laboratory. Similarly, aeronautical engineers used to build full-size airplane wings and load them up with sandbags to validate their analytical models of how these wing structures would bend and deform while loaded. In both of these instances, a prototype of a piece of a larger artifact was being built and then used to model behavior that needed to be understood as part of completing the overall design. Thus, to reiterate, engineers use prototypes to demonstrate the functionality in the real world of the actual object being designed, while we use models in the laboratory to investigate and validate behavior of a miniature of a larger artifact or of part of a large system.

We have introduced the notion of testing in discussing both models and prototypes. In design the type of testing that is often most important is *proof of concept* testing, in which it is demonstrated that a new concept, or a particular device or configuration, can be made to work in the manner in which it is being designed. When Alexander Graham Bell successfully summoned his assistant from another room with his new-fangled gadget, Bell had proven the concept of the telephone. Similarly, when John Bardeen, Walter Houser Brattain, and William Bradford Shockley had successfully controlled the flow of electrons through crystals, they proved the concept of the solid-state electronic valve, known as the transistor, that eventually replaced all of those gas-filled tubes. Laboratory demonstrations of wing structures and building connections may also be considered as proof-of-concept tests when they are used to validate a new wing structure configuration or a new kind of connection. In fact, even market surveys of new products—where samples are mailed out or stuffed into sacks in the Sunday papers—could be conceived of as proof-of-concept tests if they are designed to test the receptivity of a target market to a new product.

It is worth emphasizing that proof-of-concept tests are scientific endeavors. We are setting out reasoned and supported hypotheses that can be tested and either validated or disproved. Proofs of concept are not demonstrated simply by turning on a new artifact and seeing whether or not it "works." An experiment must be designed, often with hypotheses to be disproved if certain outcomes result. Remember, we have said that prototypes and models differ in their underlying "reasons for being" and in their testing environments. While models are tested in controlled or laboratory environments, and prototypes are tested in uncontrolled or "real world" environments, in both cases the tests are *controlled* tests. Similarly, when we're doing proof-of-concept tests, we are doing controlled experiments in which the failure to disprove a concept can be key. For example, suppose we had chosen to use Mylar containers for our new beverage product and now we're trying to design the containers to withstand shipping and handling, both in the manufacturing plant and in the store. If we think of all of the things that could go awry (e.g., stacks of shipping pallets that could topple) and analyze the mechanics of what happens in these incidents, then we might conclude that the principal design criterion is that the Mylar containers should withstand a force of X lb. We would then set up an experiment in which we could apply a force of X lb, perhaps by dropping the containers from a specifically calculated height. If the bags survived that drop, we could say that they'd likely survive shipping and handling. However, we could not absolutely guarantee survival because there is no way we could completely anticipate every conceivable thing that might happen to a beverage-filled Mylar container. On the other hand, assuming we have properly designed our experiment, if the Mylar container fails the drop test, we can then be certain that it will not survive shipping and handling, and so our concept is disproved. The National Aeronautics and Space Administration (NASA) conducted a very similar proof-of-concept test for gas-filled shock absorbers for the Mars lander. There are clearly issues of reliability involved here, and we will address some of those in Chapter 8. There are also potential issues of legal liability—for example, for how much nonstandard use of a product can a manufacturer be held responsible?—but they are equally clearly beyond our scope.

In short, prototypes, models, and proof-of-concept testing all have their place in engineering design. And while they may seem the same superficially, they are different because their intents and their test environments are different. These distinctions need to be borne in mind while deciding which of these ought to be part of the design process, and appropriate planning for them must also be done.

6.6   GENERATING AND EVALUATING IDEAS FOR THE XELA-AID DESIGN PROJECT

When we left our chicken coop teams in Chapter 5, they had come to understand many of the functions that a successful coop must fulfill, and they had developed various metrics for assessing their coop designs against project objectives. (In some cases, the metrics were not really appropriate to a one-semester project, so one of the teams had simply reduced all of their metrics down to a set of physical tests of components.) In this section we will look at some of the tools used by the teams, highlighting not only the use of the tools, but some instances where the tools could have been used more effectively.

One step that the teams had to take was to translate the various functions that had been identified into meaningful alternatives. They used morphological charts to do this. One of the teams used a single, very long morphological chart, which here is broken into two parts as Figures 6.7 and 6.8. Primary functions are indicated in boldface, while alternative subfunctions are also listed in the functions column in normal type. For example, the primary function *allow human entry/exit of coop* includes means such as *design that does not involve entering, side-hinged door(s), top hinged door(s),* and *removable roof that could be reached over.* Alternative means for keeping the coop door closed, gathered under the means *latch on door,* and itself a specific means of allowing entrance into and exit from the coop, include *hasp, hook and eye,* and *board across doors.* Note the difference in scale between these two lines of the chart. While a decision on how to keep a door latched must be made, it is certainly a different kind of decision than whether or not to have a door at all, or whether to have instead a removable roof. The difference depends on whether we are considering conceptual design or detailed design. Confusing and combining these two types of design has the unfortunate effect of cluttering up an important conceptual decision. A further problem to consider in the case of a very long chart such as that depicted in

Figure 6.7 This is the first half of a morph chart developed by one of the student teams designing a chicken coop for a Guatemalan village cooperative. Note that it is quite extensive. This completeness is a virtue in analysis, but it is likely to result in too many combinations for effective design selection.

FUNCTION	POSSIBLE MEANS					
allow for egg collection	have a sloped coop floor that sort of spirals down and eggs collect in a bin	leave where laid, allow women to pick up	ramp from each nest going to same place	nests with bucket area underneath	conveyor ramp	
protect eggs	hay in nests, if left in nests	nothing in nests, if eggs are left in nests	pads along "egg route" if not left in nests	calcium in diet (this works with others simultaneously)		
allow incubation of eggs	villagers (in greenhouse)	villagers (in coop)	allow chickens			
ventilate coop	open walls	gaps in top of solid walls, protected by overhang	"windows" in walls	fans powered by wind		
nests	one communal	several communal	individual			
allow for nest cleaning	wire bottom, so droppings fall through	removable nests that could be taken out for cleaning	nonremovable, just go in and clean	line nest with hay, replace that		
allow for chicken entry	doggie type door	use human entry door, keep closed during day	use human entry door, keep open during day	make small entry that doesn't close		
allow for waste removal	wire-mesh floor that allows waste to fall through	coop that could be lifted off base, allowing for collection	dirt floor, swept	concrete floor, swept,	wood floor, swept	potty train them
allow human entry/ exit of coop	design that does not involve entering	side-hinged door(s)	top-hinged door (s)	removable roof that could be reached over		
allow human entry/ exit of perimeter	gate in fence	no fence	climb over fence			

Figure 6.8 This is the second half of the morph chart started in Figure 6.7. In this chart, the team has offered alternatives to a number of the features that depend on previous choices, such as the size of the perimeter fence (if there is one), and the material for the roof. Once again, the team might have made better decisions had it separated the basic functions from the secondary functions. Which are basic functions and which are secondary?

FUNCTION	POSSIBLE MEANS						
Protect chickens from predators during day	perimeter fence that does not go underground	perimeter fence that does go underground	underground coop				
Protect chickens from predators during night	no added protection for night	strong coop inside perimeter for night use only	underground coop				
Protect chickens from weather	sloped roof, no walls	sloped roof on walls	flat roof, no walls	flat roof on walls			
latch on door	hasp	hook and eye	board across doors				
walls	solid walls of wood or bricks	no walls	chicken wire walls				
size of perimeter fence	high	low					
material for perimeter fence	chicken wire	wood					
material for perimeter posts	wood stakes	rebar, double or triple bound	U-posts (or T)				
roof	metal	corrugated fiberglass	corrugated metal	wood			
floor (?)	concrete	dirt					
allow for chicken feeding	wood trough, movable	wood trough, fixed to wall	"feeding room" so mess is localized	bottom of trash can or similar cylinder	individual bowls	throw food on ground	container with sides tapered
food place	carport-like roof overhang	inside coop	outside coop				
allow for chicken watering	wood trough, movable	wood trough, fixed to wall	bottom of trash can or similar cylinder	woven individual bowls	woven group bowls	concrete trough	
keep water fresh	slow running water	drain water through bottom by some sort of hole on coop floor	separate from food to prevent it from getting wet				

Figures 6.7 and 6.8 is that the number of possible combinations can become very high. Recall that in Section 6.3 we saw that our beverage container design, which had only a few functions, had 864 possible outcomes. For this coop, the total number of outcomes is overwhelming. Clearly, some strategy for grouping and organizing the functions and the resultant design alternatives is called for.

The second team did break the functions into general areas, as shown in Figures 6.9 (a)-(d). In this case the team organized the functional decisions in terms of: *safety of inhabitants, egg production, growth of chicks,* and *maintenance of adult birds.* This allows each morph chart to have only three or four functions, and so it results in more manageable alternatives. In some of the cases, the alternatives that are considered for one of the functional areas may also apply to others, which further constrains the number of alternatives that must ultimately be considered. For example, *egg incubation* and *brooding of chicks* are clearly related in terms of the role of the *broody hens* insofar as it is unlikely that we would use both *egg incubation* and *broody hens* to raise the young chicks.

In terms of generating alternatives, then, it is useful to note that the teams could have created a large number of alternatives from various combinations. It is equally useful to note that many, if not most, of these are actually minor variants that are not conceptually different. In fact, the primary conceptual alternatives that the teams developed might be characterized in terms of three basic design sets: cages for containing the birds while they grow and mature (much like those used in high-production chicken farms in the U.S.); a fenced-in area which encloses a small building for the chicks and nests; and a fenced-in area containing separate areas for nests (egg production) and for roosting. Different versions of each of these alternatives were explored by both teams. Within these alternatives, the teams explored a number of variations, including greater and lesser dependence on villager labor, different roofing and fencing ideas, and, ultimately, different construction materials. Within their conceptual designs, each team also considered several more detailed design variations.

Both teams attempted to use simplified versions of the evaluation instruments proposed earlier in this chapter, using what could probably best be described as a verbal version of the best of class method in which alternatives are ranked and then weighted by the relevant objectives. This resulted in some interesting but quite different results for the two teams.

The outcome of the process led one of the teams to select a coop design in which all the available space would be used for a building enclosing cages and a limited number of nestboxes. Figures 6.10 and 6.11 show sketches for this design choice. The cages were made largely of rebar and chicken wire, with supporting posts of concrete block. These cages were designed and built to stack two levels high. This design allowed the team to use all the available space for production and would, if constructed, result in a predicted capacity of 64 chickens and 12 or more chicks. However, the team predicted that the cost of this design would be quite high.

The other team elected to build a $6 \times 3 \times 4$ (ft) coop structure surrounded by a wire fence. Figure 6.12 shows a sketch of the structure, which was to be built of locally available concrete blocks. It would be divided into two halves: one for nests, the other for roosting. The structure would be surrounded by a gated fence

Figure 6.9 These are morph charts from another team that designed a chicken coop for the Guatemalan village cooperative. Note that this team has organized the functions into several categories to simplify its decision making. Notice also that many of the "functions" could be easily reworked into the standard "verb-object" syntax that we discussed in Chapter 5.

(a) Safety of inhabitants

Functions	Means			
Predator exclusion	Metal skirt on posts	Concrete floor, walls	Sand or gravel floor	Extend fence below ground level
Waste removal	Waste pit	Dropping boards	Waste falls through sunfloor	Treat waste with lime
Climate control	Chicken wire walls	Burlap-covered chicken wire	Windows	Sheet metal roof
Prevention of intra-coop violence	Separate chicks from adults	Separate turkeys, chickens	Round off interior corners	

(b) Egg Production

Functions	Means		
Egg collection	Dark nest boxes	Communal nests	Cages
Egg incubation	Broody hens	Box with heat lamp	In humans' homes
Separation of infertile (eating) eggs	By hand during incubation (candling)		

(c) Growth of chicks

Functions	Means		
Brood chicks	Broody hens	Box with heat lamp	In humans' homes
Water chicks	Troughs	Fountains (inverted jar on tray)	Collect rain into troughs or basins
Feed chicks	Communal troughs filled daily by hand		

(d) Maintenance of adult birds

Functions	Means			
Promote growth	Allow access to insects	Mix ground meat into feed (byproducts)	Exercise area	Ground seashells in food
Water	Troughs	Fountains (inverted jar on tray)	Collect rain into troughs or basins	
Feed	Communal troughs filled daily by hand			

Figure 6.10 This is a sketch of the outer structure for a chicken coop design proposed by one of the the student teams. The team has chosen to enclose the entire space and then fill that area with cages and nesting areas, following the example of U.S. poultry factories. The size of the beams used to support the structure suggests that the outer shell can also be used as a railroad bridge if properly stabilized.

that would be buried one foot underground to keep out predators. This design was predicted to support 50 chickens and be less expensive than other teams'. It was not, however, very cheap.

An interesting question that immediately comes to mind is why two teams in the same class would come to such different conceptual designs for the same client. The answer can be seen in Figures 3.9 and 3.10, wherein the teams attached weights to the client's objectives. One team rated increased production with a weight of more than half of the overall objectives (55%), while the other had a rating of only 11% for this, focusing instead on matters such as the safety of inhabitants (28%), survival in the Guatemalan climate (25%), and ease of repair (14%). As a result, one team placed a very high premium on the number of chickens that could be supported, even at the expense of higher initial costs and greater design complexity, while the other team focused on other matters, such as ease of repair or stability against natural forces. In Chapter 7 we will see which of the conceptual designs was ultimately adopted by the villagers and Xela-Aid.

It is instructive to note how the teams sought to model and test their design choices. Both teams faced a common problem that designers must overcome, namely, both wished to confirm their designs but neither could build and operate a functioning chicken coop within their time and budget constraints. In the case of the cage team, the key concept that it wanted to establish was that safe usable cages could be constructed with available materials. As a result, the team

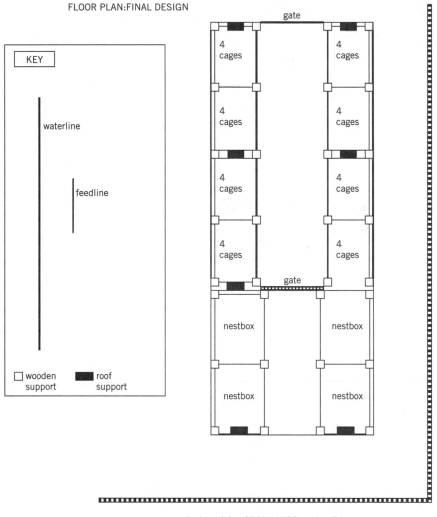

cages are duplex, giving 64 birds, (32 per level)

Figure 6.11 This sketch lays out the internal structure of the proposed cage-coop depicted in Figure 6.10. Note that the team has determined a means to utilize most of the space.

first constructed a stack of two cages, using concrete blocks, rebar, and wire. It found, somewhat to its dismay, that cages constructed according to this way would not only be very heavy; they would also be unstable unless lateral bracing were provided—and that would be quite difficult and complicated. (The team's basic hypothesis was that the cages would resist moderate lateral forces, defined in terms of one person pushing with moderate force on the structure. In fact, such forces caused a great deal of "sway" and thus served to *disprove* the hypothesis.) As a result, the team's final design substituted a wood frame as support for the cages of rebar and chicken wire. The team also learned that actually building the cages as originally designed would be very complicated and time

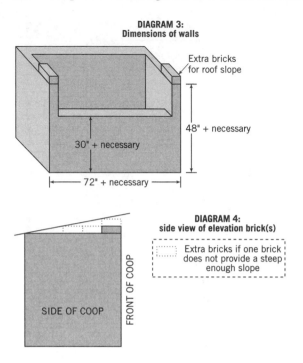

DIAGRAM 3:
Dimensions of walls

Extra bricks
for roof slope

48" + necessary

30" + necessary

72" + necessary

DIAGRAM 4:
side view of elevation brick(s)

Extra bricks if one brick
does not provide a steep
enough slope

SIDE OF COOP

FRONT OF COOP

Figure 6.12 Another student team's approach to the coop problem was to fence in the entire available space, and use a smaller building within that fence to provide shelter and nesting areas for the birds. This structure has a lower wall in the front that acts to support nest-boxes and egg drawers. It is not obvious that this function could not be met by other means in ways that would allow for easier cleaning and ventilation.

consuming, so several simplifications were needed. For this, the team timed the construction of a complete cage when materials were available. Once again, its hypothesis that the cages could be built as designed within the time specified was disproved. These test results—which failed to correspond to its expectations and led the team to modify and improve its design— provide a good example of how testing can be used to identify both design shortcomings and possible fixes.

The second team also wished to test certain components of its design prior to completing it and delivering it to the client. In this case, the ability to build a safe and strong fence using materials known to be locally available was a key element in the design. To this end, the team considered several alternatives, including using single pieces of rebar for the posts, doubled pieces, and triple pieces. There is an obvious cost advantage to fewer pieces, as well as greater ease in construction. Unfortunately, the team learned through experimentation that only triple rebar, wired together at intervals of 18 in or less, could support both the weight of the fence and reasonable estimates of the lateral forces. Once again, their experiments, while initially perhaps disappointing, led the team to a better design—and that is the ultimate purpose of proof-of-concept testing.

6.7 MANAGING THE GENERATION AND SELECTION OF DESIGN ALTERNATIVES

Throughout this chapter (as well as in previous discussions of brainstorming and other team discussions), we have emphasized the importance of team dynamics and, especially, respect for the ideas of others. We have done this because, in this phase of a design project, the interpersonal aspects of how a team uses the formal tools and techniques are as important as the tools and techniques themselves. Having said that, however, we think it is also important to consider in these processes the roles of some of the more formal management tools discussed in Chapter 4.

Recall that in Chapter 4 we introduced tools for task management (the WBS), scheduling activities (Gantt charts and activity networks), budgeting, and monitoring and controlling progress (the PCM). We now briefly look at each of these in terms of how they can help with generating and evaluating design alternatives, and with proof-of-concept testing and prototyping.

Before turning to the specifics, however, we make some general observations. The first is that the relative space devoted in this book to describing each of the activities is not an accurate estimate of the actual time that a team will need to conduct them. It is likely that the idea generation activities can and will proceed relatively quickly, while demonstrating proofs of concept may take much more time. This is a natural consequence of a team's prior investment of thought and study in the problem. While formal alternative generation may begin at a specific time, most of the team members will have been considering alternatives throughout the project. Evaluation of the design alternatives, on the other hand, requires that all of the alternatives be laid out systematically, and so it cannot be distributed over the life of the project in the same way. Proof-of-concept testing of an idea is likely to be a "time sink" for many design teams since it often requires the team to apply skills they might not know they have and to order parts and tools. Thus, concept testing may require several iterations to "make it work." All of this should cause a team to revisit its previous management tools and plans, both to update them and to determine whether or not all of its ambitious goals can actually be realized.

A second point is that the activities described in this chapter offer very real opportunities for a design team to consider how it allocates responsibility for these activities among its members. To be fully effective, brainstorming and other creative activities demand that the entire team be present. Building a physical model, on the other hand, may be best accomplished by a subset of the team. Hard feelings are almost certain to result if this staffing problem is not handled properly, so the entire project can be damaged.

A final general point is that design projects, whether in academia or out in the "real world," almost always proceed under the pressure of fixed deadlines and other competing activities. This places a premium on time management: Adjustments should be made early, that is, as soon as information permits. There is simply no substitute for having the team review its work plan at regular intervals and make changes as needed.

6.7.1 Task Management

At the start of a design project, a team may well have been sufficiently unfamiliar with the particular design problem being solved (and perhaps design processes in general) that it may have constructed a *pro forma* WBS, that is, the team may have "filled in the blanks" without seriously thinking through the implications and consequences of the time assignments being made. However, by this point in a project, the team should have a much better understanding of its problem and of the tasks that it must undertake to complete the project. For this reason, the team should, at a minimum, review its WBS before it generates and evaluates design alternatives, and before proceeding with proof-of-concept testing and prototyping. Testing and prototyping in particular will only be useful to a team *after* design concepts have been generated and successfully evaluated. Among the tasks the team needs to consider are:

- selecting functions and actions of the prototype or physical model,
- acquiring parts or special materials needed to build the prototype,
- building the prototype,
- developing a test protocol,
- conducting the test, and
- revising and reworking when the prototype doesn't work initially—and it never does!

It is also often the case that teams find out at this late time in the project that their prototype or proof of concept will not be as complete as they had hoped originally. This ugly fact must be squarely faced and considered in terms that will allow either the client or a subsequent designer to build on the work done that has been completed. The results may be useless without this perspective.

6.7.2   Scheduling

Just as the tasks of the project may require review and revision during the alternative generation, alternative evaluation, and testing stages, the schedule under which they can be accomplished should also be reviewed. Depending on the time remaining to complete the project and the complexity of the tasks yet to be done, one of the methods proposed in Chapter 4 should be applied.

In many cases, teams that have been using a computer-generated activity net or some other standard project tool can modify their remaining activities and schedules quite easily. An important point for the team to note is that it must review not only its responsibilities under this project, but also any other specific commitments that its members may have. Team members need to thoroughly think through how their personal commitments—both on the job and off, both schoolwork and extracurricular—might affect their ability to devote time to the project being scheduled. Most software packages permit users to specify availability calendars for each team member. This calendar should be updated as new information (e.g., new projects) becomes available. Most teams that then recalculate their schedule on the computer are both surprised and dismayed. (Co-authors of books have similar experiences!) However, even if this revised schedule is discouraging, it should not be allowed to become an excuse for disregarding the schedule. Rather, it should be used as a reminder of the need for discipline in adhering to a schedule.

Teams that have been using a simpler team calendar for planning activities now must pay a price for their previous convenience. Clearly, a simple calendar cannot calculate anything, so the team must, as a group, revisit its commitments and their implications. This may cause the team to set target dates that require some number of "all-nighters" or other unplanned or undesirable activities. If that is the case, the team has to "sweat the details" of scheduling, probably at every team meeting.

With either approach, the team will have to consider the logical and temporal relationships between activities. It is difficult, if not impossible, to assemble a prototype if the parts haven't arrived. Parts are unlikely to arrive if they haven't been ordered. Parts can't be ordered if they haven't been selected. Ignoring these seemingly obvious orderings of events has been the pitfall of many design teams, both in professional firms and in academic projects. They can be avoided with planning and (remember Keynes!) "a little clear thinking" ahead.

6.7.3 Budgeting

Closely related to the above discussion is the need for a team to create and use an appropriate budget. In many projects, there is an upper limit on the financial resources available to that team. It is important that this be considered before the team develops all of their metrics or their methods of proving the concept. It is important to keep in mind here that the budget for building a prototype or conducting a proof-of-concept experiment should not be confused with the budget and costs of building the final product. That distinction will be explored in Chapter 8, but for now it is important to realize that design teams are almost always faced with resource limits in establishing their design choices. In many cases, teams will not be able to prove all of their key ideas within their available budget or resources. In such cases, it is necessary to set priorities and decide which concepts must be demonstrated by building prototypes or other physical models and which concepts can be established either by analysis or by reference to appropriate authority.

Once the team has agreed on how to spend the resources available to demonstrate its design choices or prove its concepts, the team should agree on a simple process for approving and making expenditures. In some cases, a team's parent organization will have explicit formal procedures (e.g., purchase orders, reimbursement forms, etc.), but even in these cases, someone on the team should have the responsibility for tracking and recording all expenditures on behalf of the team. This helps avoid later problems with budget overruns or disapproved expenditures.

6.7.4 Monitoring and Controlling Progress

All of the foregoing discussion of this section has presumed that the team has a mechanism for monitoring and tracking progress. We discussed appropriate tools for this in Chapter 4, but it is worthwhile to take a moment and discuss briefly how these tools can now be translated into team-level success.

Probably the single most important element in monitoring and tracking a team's activities relative to its plan is explicit, formal communication about these topics at team meetings. Every team meeting should include a brief review of the recent and near-future elements of the work plan, a discussion of progress to date, and consideration of any changes to the time allowed for activities.

Team reviews of the team's recent, current, and future activities allow each member of the team to see what is being done, to identify her responsibilities and commitments, and to report on her work to date. It is clearly a good thing to disseminate information this way within the team. It also tends to encourage members who may have been doing less work to do more, if only because they want to avoid having "nothing to report." Such a review of activities should be a scheduled part of the team-meeting agenda, either at the outset as a means of bringing everyone up to speed on team progress, or as the final agenda item, as a way of stimulating further work.

Discussion of progress to date on the tasks should, wherever possible, be focused on concrete or quantifiable goals and targets, avoiding general remarks such as, "We're coming along on that." Ideally, tasks can be examined in terms of specific deliverables or milestones. Alternatively, it may be useful to work toward a specific percentage of work completed on a task. It may seem artificial initially to force a responsible team member to indicate that a task is 25, 50, or 75 percent complete, but it can quickly become a group norm that allows members to gauge their own progress and to request help when it's needed.

After discussing the tasks to be done and the progress made to date, the team should examine the implications of its current situation in the light of tasks yet to be done. This should almost certainly be done at or near the end of a meeting, since new tasks or modifications may well have emerged previously during the meeting. Team leaders should allocate some time at the end of each meeting for this, preferably as part of a final agenda item that includes progress assessment. It is really a mistake to try to update schedules and timelines as everyone is picking up her coat and heading for the door! An important element of this final updating is that someone on the team should be responsible for sending out (perhaps by electronic mail) a revised schedule and percent-complete matrix (PCM) that reflects the project status as of the last meeting. This allows members to review the commitments they made in the meeting. It also affords the team a record of their progress as the project unfolds.

At first, all of these management tasks may seem to "get in the way" of the engineering and the design. While managing the project may seem like a lot of work, experience shows that unmanaged teams do even more work. In fact, unmanaged teams usually do things more than once. There is an old saying to the effect that teams rarely accomplish more than they plan on. This is certainly true of most design teams.

6.8   NOTES

Section 6.2: The address of the USPTO's Web site is www.uspto.gov. Another often-used Web site is www.ibm.com/patents. Group methods of idea generation are explored and described in (Shah 1998) and synectics are described in (Cross 1994). Approaches to creativity and analogical thinking in a group setting are described in (Hays 1992).

Section 6.3: Zwicki (1948) originated the idea of a morphological chart. Examples of morph charts can be found in (Cross 1994), (Jones 1992), and (Hubke 1988).

Section 6.4: Our dictionary is the *Webster's Ninth New Collegiate Dictionary* (Mish 1983).

Section 6.6: The results from the Xela-Aid chicken coop design project are taken from final reports ((Gutierrez et al. 1997) and (Connor et al. 1997)) submitted during the Spring 1997 offering of Harvey Mudd College's freshman design course, E4: Introduction to Engineering Design.

6.9 EXERCISES

6.1 Explain what is meant by the term "design space" and discuss how the size of the design space might affect a designer's approach to an engineering design problem.

6.2 Using the functions developed in Exercise 5.4, develop a morphological chart for the portable electric guitar.

6.3 Organize and apply a process for selecting means for realizing the design of the portable electric guitar.

6.4 Using Web-based patent lists (identified in Section 6.8), develop a list of patents that are applicable to the portable electric guitar.

6.5 Using the functions developed in Exercise 5.6, develop a morphological chart for the rain forest project.

6.6 Organize and apply a process for selecting means for realizing the design of the rain forest project.

6.7 Using Web-based patent lists (identified in Section 6.8), develop a list of patents that are applicable to the rain forest project.

6.8 Describe an acceptable proof of concept for the rain forest project. Would a prototype be appropriate for this project? If so, what would be the nature of such a prototype?

Chapter 7

Reporting the Outcome

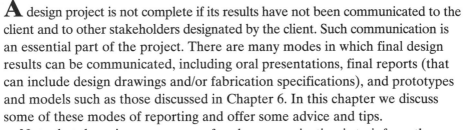

How do we let our client know about the solutions?

A design project is not complete if its results have not been communicated to the client and to other stakeholders designated by the client. Such communication is an essential part of the project. There are many modes in which final design results can be communicated, including oral presentations, final reports (that can include design drawings and/or fabrication specifications), and prototypes and models such as those discussed in Chapter 6. In this chapter we discuss some of these modes of reporting and offer some advice and tips.

Note that the primary purpose of such communication is to inform the client *about the design*, including explanations of how this design was chosen over other competing design alternatives, and why. It is most important to convey the *results* of the design process; the client is unlikely want to know the history of the project or about the internal workings of the design team. Thus, final reports and presentations are not chronologies of the team's work. Rather, and ideally, they are lucid descriptions of the *outcomes* of the design process.

7.1 THE PROJECT REPORT: WRITING FOR THE CLIENT, NOT FOR HISTORY

As just noted, the purpose of the final or project report is to communicate with the client in terms that ensure the client's thoughtful acceptance of a team's design choices. The client's interests demand a clear presentation of the design problem, including analyses of the needs to be met, the alternatives considered, the bases on which decisions were made, and, of course, the decisions that were taken. The results should be summarized in clear, understandable language. One practice that we recommend to support clarity is consigning highly detailed or technical materials to appendices placed at the end of the report. In fact, it is not unusual, and in large public works projects it is the norm, to move all of the technical and other supporting materials to separate volumes. This is an especially important approach when the client and the principal stakeholders are not engineers or technical managers, but perhaps members of the general public.

The process of writing a technical report, like so much of what we have described about doing design, can be managed and controlled with a structured approach. In fact, there are striking similarities between the design process and report writing, especially in their early, conceptual stages. In both instances it is very important to clearly delineate objectives, for both the designed object and for the project report. In both instances it is very important to understand the "market," that is, to understand the user needs for the design and the intended audience of the final report. In both instances it is very important to be reflective and analytical and to recognize that analysis isn't limited to applying known formulas. For the design process, we described several tools that could be used to support our thinking. Similarly, writing is also an analytical thinking process.

As with the design process, the use of structure is not intended to displace initiative or creativity. Rather, we find that structure can be helpful as a way of learning how to construct an organized report of the design results. In this case, one structured process that a design team might follow would include the following steps:

- construct a rough outline of the overall structure of the report;
- review that outline within the team and with the team's managers or, in case of an academic project, with the faculty advisor;
- construct a topic sentence outline and review it within the team;
- distribute individual writing assignments and write an initial draft;
- solicit reviews of the initial draft from managers and advisors;
- revise and rewrite the draft in response to reviews of the initial draft; and
- prepare the final version of the report and present it to the client.

We now discuss each step in greater detail.

7.1.1   The Rough Outline: The Structure of the Final Report

Only a fool would start building a house or an office building without first analyzing the structure being built and then organizing the building process. Yet many people sit down to prepare a technical report and immediately begin writing, without trying to lay out in advance all of the ideas and issues that need to be addressed and without considering how these ideas and issues relate to each other. One result of such unplanned report writing is that the report turns into a project history, or worse, it sounds like a "What I Did Last Summer" essay: First we talked to the client, then we went to the library, then we did research, then we did tests, etc., etc., etc. While technical reports may not be as complex as high-rise office buildings or airplanes, they are still complicated enough that they cannot simply be written as simple chronologies. They must be planned!

The first step we need take to write a good project or final report is building a good rough outline that documents the large-scale structure of the final report. That is, we should identify the major sections into which the report should be divided. Some of the sections typically found in a technical report are:

- Abstract
- Executive summary
- Introduction and overview
- Analysis of the problem, including relevant prior work or research
- Design alternatives considered
- Evaluation of design alternatives and basis for design selection
- Results of the alternatives analysis and design selection
- Supporting materials, often set out in appendices, including:
  - Drawings and details
  - Fabrication specifications
  - Supporting calculations or modeling results
  - Other materials that the client may require

This outline may look like a table of contents, and it should, because the final report of an engineering or design project must be organized in a way that a reader can go to any particular section and see it as a clear and coherent stand-alone document. It is not that we think things should be taken out of context. Rather, it is that we expect each major section of a report to make sense all by itself; that is, it should tell a complete story about some aspect of the design project and its results.

Having identified a rough outline as the starting point for a final report, it is appropriate to wonder *when* should that outline be prepared. Indeed, when should the final report be written? It seems "obvious" that we can't write a *final* report before we have completed our work and identified and articulated a final design. On the other hand, as with the design process, it is very helpful to have some idea of where we're going with a final report so that we can begin to organize and assemble it along the way. It can be very helpful to develop a general structure for the final report early in the project. This enables us to track and appropriately file or label key documents from the project (e.g., research memoranda, drawings, objectives trees) according to whether their contents would appear in the final report. Thinking about the report early on also places a premium on thinking about the project's *deliverables*, that is, those things that the team is contracted to deliver to the client during the entire project. Beginning to organize the final report early on may make the final stages or endgame of the project much less stressful, simply because there will be fewer last-minute things to identify, create, edit, and so on, so they can be inserted into the final report.

7.1.2   The Topic Sentence Outline: Each Entry is a Paragraph

It is a cardinal rule of writing that each and every paragraph of a piece should have a topic sentence that indicates the intent or thesis of that paragraph. Once the rough outline of a report has been established, it is usually quite useful to build a corresponding detailed *topic sentence outline* (TSO) that will identify the themes or topics that, collectively, tell the story told within each section of the overall outline of the report. Thus, if a topic is identified by an entry in the TSO, we can assume that there is a paragraph in which that topic is covered.

The format of a TSO enables us to assess the completeness and follow the logic of the argument or story of each section we are drafting, as well as of the report as a whole. Consider the case where there is only one entry in a TSO for something that we consider important, say, the evaluation of alternatives. One implication of this is that the final report will have only one paragraph devoted to this topic. Since the evaluation of alternatives is a central issue in design, it is very likely that there should be entries on a number of aspects, including the methods of evaluating design options, the results of the evaluation, key insights learned from the evaluation, the interpretation of numerical results—especially for closely scored alternatives, and the outcome of the process. Thus, a quick examination of a TSO can clearly show us that a proposed report is not going to address all of the issues that it should.

For the same reason, of course, TSOs help us identify appropriate cross-references that we may need to make between subsections and sections as different aspects of the same idea or issue are addressed in different contexts. Taking advantage of the format of a TSO, we also find them useful for eliminating needless duplication because it usually makes it much easier to spot repeated topics or ideas. In Section 7.4 we show an excerpt from a TSO that demonstrates some of these points.

There's little doubt that writing in this way is a hard discipline, but TSOs do provide a number of advantages to a design team. One is that a TSO forces the team to agree on the topics to be covered in each section. It quickly becomes clear, then, if a section is too short for the material, or if one of the co-authors (or team members) is "poaching" on another section that was agreed upon in the rough outline. A further advantage of a good TSO is that it makes it much easier for team members to take over for one another if something comes up to prevent a "designated writer" from actually writing. For example, a team member may suddenly find that the prototype is not working as planned and she thus needs to do some more work on it. TSOs also make life easier for the team's report editor (see the next section) to begin to develop and use a single voice.

One final note on topic sentence outlines. Notwithstanding the definition of the abbreviation TSO, the entries in a TSO do not have to be complete sentences. However, they should be complete enough that the content of a topic sentence is clear and unambiguous.

7.1.3   The First Draft: Turning Several Voices Into One

One advantage of agreed-upon rough and topic sentence outlines is that having such common structures of the report's contents allow teams members to write in parallel or simultaneously. However, this advantage carries a number of price tags, most notably that corralling the efforts of several writers into a single, lucid, coherent document is something like herding a bunch of cats. As a matter of fact, the more writers, the greater the need for a single, authoritative editor. Thus, one design team member should enjoy the rights, privileges, and responsibilities pertaining to being the editor. Further, this designation should be done by the team as soon as the planning of the report begins, hopefully at or near the onset of the project.

The editor's role is to ensure that the report has the characteristics of continuity, consistency, accuracy, and a single voice or style. *Continuity* means that topics and sections follow a logical sequence that reflects the structure of the ideas in the rough outline and the TSO. *Consistency* means that the report uses common terminology, abbreviations and acronyms, notation, units, similar reasoning styles, and so on, throughout the report and all of its appendices. It also means, for example, that the team's objectives tree, weighted objectives tree, pairwise comparison chart, and evaluation matrix all have same elements; if not, any discrepancies should be explicitly noted and addressed.

Accuracy means that any calculations, experiments, measurements, or other technical work are done to appropriate professional standards and current best practices. Such standards and practices are often specified in contracts between a design team and its client(s). They typically provide that stated results and conclusions must be supported by the team's prior work. Accuracy, as well as intellectual honesty, require that technical reports should not make unsupported claims. In the closing moments of a project, there is often a temptation to add to a final report something that wasn't really done well or completely. Needless to say, this is a temptation that should be avoided.

The *voice* or style of a report reflects the way in which a report "speaks" to the reader, in ways very similar to how people literally speak to one another. It is very important, even essential, that a technical or final report *speaks with a single voice*. This mandate has several facets. The first is that the report has to read (or "sound") like it was written by one person, even when its sections were written by members of a very large team. In the same way that a president of the United States uses several speechwriters but still sounds like the same, familiar person, so a technical report must read in a single voice. Ensuring that single voice is one of the most important duties of the editor. Further, that voice has to be, typically, rather more formal and impersonal than the voice of this book. Technical reports are not personal documents, so they cannot sound familiar or idiosyncratic. Also, that single voice can either be active or passive as modern practice renders both acceptable. What is important is that the voice of the report be the same in the opening Abstract and the closing Conclusions.

Clearly, there are potentially serious implications for the team dynamics of the writing process. Team members have to be comfortable surrendering control of pieces they have written, and they have to be willing to let the editor do her job. We will discuss aspects of the team dynamics of report writing in Section 7.5.1.

7.1.4   The Final Version of the Final Report: Ready for Prime Time

A good review process can ensure that a draft final report gets thoughtful reconsideration and meaningful revision. Draft reports can really benefit from careful readings and reviews by team members, managers, client representatives or liaisons, faculty advisors, and others whose connection with the project may be very distant. This means that as we are trying to wrap up our project report, we need to incorporate the reviewers' comments, suggestions, and any resulting changes into a final, high-quality document. There are a few more points to keep in mind.

We'd like our final report to be *polished*. It should be a professionally done report. We do not simply mean that it has to have glossy covers, fancy type and graphics, and an expensive binding. Instead, it means that the report will be clearly organized, easy to read and understand, and that its graphics and figures are also clear and readily interpreted. The report should also be of reproducible quality because it is quite likely to be photocopied and distributed within the client's organization, as well as possibly to other individuals, groups, or agencies beyond.

We should keep in mind, as foreshadowed by our last cautionary note, that the report may go to a very diverse audience, not simply to peers. Thus, while the editor needs to ensure that the report speaks with a single voice to an anticipated audience, she should try as much as possible to also ensure that the report can be read and understood by readers that may have different skill levels or backgrounds than either the design team or the client. An *executive summary* is one way to address readers who may not have the time or interest to read all of the details of the entire project.

Finally, the final report will be reviewed and then used by our client(s), who will, we hope, adopt our design. This suggests that the report and its appendices and other supporting materials be sufficiently detailed and complete to stand alone as the final documentation of the designs we are proposing. We will have more to say about this in Section 7.3.

7.2   ORAL PRESENTATIONS: TELLING A CROWD WHAT'S BEEN DONE

Most design projects call for a number of meetings and presentations to clients, users, and technical reviewers. These presentations may be made early in the design project, or even before the award of a contract to do the design job, and can occur often over the life of the project. Early presentations may include briefings that focus on the design team's ability to understand and do the job in the hope of winning the design job in a competitive environment. During the project, the team may be called upon to present their understanding of the project (e.g., the client's needs, the artifact's functions, etc.), the alternatives under consideration and the team's plan for selecting one, or simply their progress toward completing the project. After a design alternative has been selected by the team, the team is often asked to undertake a design review before a technically skilled audience for the purposes of assessing the design, identifying possible problems, and suggesting alternate solutions or approaches. At the end of a project, design teams usually report on the overall project to the client and to other stakeholders and interested parties.

Because of the variety of presentations and meetings that a team may be called upon to make, it is impossible to examine each of them in detail. However, we do note that there are elements common to most of them, so it may be useful to offer some general guidelines for preparing and conducting presentations. We will do that in Section 7.2.1, after which we discuss the design review, a particularly important kind of presentation.

7.2.1   Presentations to Clients and Users

In general, there are several key notions that underlie all successful design presentations. Foremost among these are the need to understand the audience, prepare thoroughly for the presentation, practice the presentation, and develop and use supporting materials appropriate to the presentation.

7.2.1.1   *Knowing the Audience: Who's Listening?*

There are many types of audiences for design briefings and presentations. For example, some projects call for the work to be reviewed at certain points by technical managers. Others are concerned with the management implications of a design. Still others may be concerned with how a design will be manufactured. Consider, for example, our new beverage container. We might be called upon to present our design work to logistics managers who are concerned with how the containers will be shipped to warehouses around the country. We might also be asked to present our design alternatives to marketing professionals concerned with the effect the designs might have on establishing brand identity. A briefing of manufacturing managers might be scheduled in order to identify any special production needs of design alternatives. In each case, the design team must understand the audience in terms of several factors, including:

- Different members of the audience will have differing *levels of interest* in the project. While we can assume that most attendees at a meeting are interested in at least some aspect of the project, it may be the case that some, or even most, are only interested in particular dimensions of the project. If we can gauge these particulars in advance, the presentation can be targeted to those interests. A team usually can learn this simply by asking the organizer of the meeting, whether that is the client, some users, or some other stakeholders.

- Different members of the audience are likely to have different *levels of understanding of the problem*. It is usually best to assume that the audience will have a less complete understanding of a problem than the design team, so at least some of the presentation should be allocated to review the nature of the design problem being solved.

- It is very important to understand the *technical skill levels* of the audience. For example, a presentation to a group of practicing engineers will usually be quite different than a presentation to marketing professionals. If the audience includes several kinds of professionals, then it may be necessary to define and clarify various terms in the presentation. It is critical that the presenters identify and use the "languages" that audience members understand.

- Another important aspect of the audience, and of presentations in general, is the *amount of time available* for the presentation. It may be the case that some parts of an audience will be attending only part of the presentation. Knowing this allows a team to ensure that relevant information is presented within that time window.

Once the audience is known and understood, a team can begin to plan and develop its presentation, tailoring it to that audience. Like any other deliverable, the presentation must be organized and appropriately structured. The first step in developing a presentation is the articulation of a rough outline; the second is the formulation of a detailed outline; and the third is the preparation of the proper supporting materials, such as visual aids or physical models.

7.2.1.2   The Presentation Outline

Just as with the final report discussed in Section 7.1, a presentation should have a clear structure. We achieve this structure by developing a rough outline, analogous to the rough outline used in writing a technical report. The structure of the presentation then guides the team as it prepares supporting commentary and discussion. Like the final report, a presentation should have a logical and understandable organization. And because a design presentation is not a movie or a novel, the team should avoid trying to find "surprise endings." An outline for a sample presentation might include the following elements:

- *A title slide*, giving the name of the project and the team or organization responsible for the project.
- *An overview of the presentation*, allowing the audience to see the direction that the presentation will take.
- *A statement of the problem*, including the initial problem as given by the client and some indication of how that problem statement was changed as the team came to understand it.
- *Background material on the problem*, including relevant prior work and other materials developed through team research.
- *The key objectives of the client and users*, probably including the information in the top level or two of the objectives tree.
- *Functions that the design must fulfill*, focusing on basic functions, but possibly including key issues related to unwanted secondary functions.
- *Design alternatives*, particularly those that were considered seriously in the evaluation stage.
- *Highlights of the evaluation procedure and its results*, including key metrics or weighted objectives that bear heavily on the outcome.

- *The selected design*, explaining why this design was chosen.
- *Features of the design*, including aspects that make it superior to other alternatives and any novel or unique features.
- *The results of proof of concept testing*, especially if the audience is composed of technical professionals for whom this is likely to be of great interest.
- *A demonstration of the prototype*, assuming that a prototype was developed and that it can be shown. Videotapes or still photos may also be appropriate here.
- *The conclusion(s)*, including the identification of any future work that remains to be done.

There may not always be enough time to include all of these elements in a talk or presentation, so a team may need to limit or exclude some of them. This decision will also depend at least in part on the nature of the audience.

Once the rough outline has been articulated, a topic sentence outline for the presentation should also be developed. This is important both to ensure that the team understands the point that is being made at each point in the presentation and to develop corresponding bullets or similar entries in the supporting visual aids, e.g., slides or transparencies. Generally, bullets should correspond to entries in the topic sentence outline, although in many cases there will be points to be made for which there is no corresponding entry on a slide or transparency.

Just as developing the topic sentence outline for a report appears at first to be cumbersome, the detailed outline for the presentation may seem like a great deal of work. Ironically, team members having the most experience in public speaking may be most resistant to these formalizations. This is likely the result of their already having internalized a similar method of preparation. However, since presentations represent entire teams, all members of a team should review the structure and details of the presentation, and this review in turn requires that a detailed outline be constructed.

7.2.1.3 *Practice Makes Perfect, Maybe . . .*

Presenters and speechmakers are usually effective because they have extensive experience. They have given many speeches and made many presentations, as a result of which they have identified styles and approaches that work for them. Design teams cannot conjure up or create such real-world experience, but they can practice a presentation often enough to gain some of the same confidence that experience breeds. To be effective, speakers typically need to practice their parts in a presentation alone, then in front of others, including in a "rump" audience at least some people who are not already familiar with the topic.

Another important element of an effective presentation is that a speaker uses words and phrases that are natural to him. Similarly, each of us normally has an everyday manner of speech with which we're quite comfortable. While developing a speaking style, however, we have to keep in mind that ultimately we want to speak the language of our audience, and that we want to maintain an appropriately professional tone. Thus, when practicing alone, it is useful for a presenter to try saying the key points in several different ways as a means of identifying and adopting new speech patterns. Then, as we find some new styles that work, we should repeat them often enough to feel some ownership.

Practice sessions, whether solitary or with others, should be timed and done under conditions that come as close as possible to the actual environment. Inexperienced speakers typically have unrealistic views of how long their talk will last, and they also have trouble setting the right pace, going either too fast or too slow. Thus, timing the presentation—even setting a clock in front of the presenter—can be very helpful here. If slides (or transparencies or a computer) are to be used in the actual presentation, then slides (or transparencies or a computer) should be used in practice.

The team should decide in advance how to handle questions that may arise. This should be discussed with the client or the sponsor of the presentation before the team has finished practicing. There are several options available for handling questions that are asked during a talk, including deferring them to the end of the talk, answering them as they arise, or limiting questions during the presentation to clarifications of facts while deferring others until later. The nature of the presentation and the audience will determine which of these is most appropriate, but the audience should be told of that choice at the start of the presentation. When responding to questions, it is often useful for a speaker to repeat the question, particularly when there is a large audience present or if the question is unclear. The presenter or the team leader should refer questions to the appropriate team members for answers. If a question is unclear, the team should seek to clarify it before trying to answer it. Just as with the presentation itself, the team should practice handling questions that it thinks might arise.

There are several ways of practicing talks and preparing for questions, including:

- generate a list of questions that might arise and prepare for them;
- prepare in advance supporting materials for points that are likely to arise (e.g., slides, computer results, statistical charts, etc.); and
- don't be afraid to say "I don't know" or "We didn't consider that." This is a really important point. To be caught *pretending* to know both undermines the presenter's (and the team's) credibility and invites severe embarrassment.

A final note about selection of speakers is in order. Depending on the nature of the presentation and the project, the team may want to have all members speak (for example, as a course requirement), it may want to encourage less experienced members to speak in order to gain valuable experience and confidence, or the team may want to tap its most skilled and confident members. As with so many of the presentation decisions, choosing a "batting order" will depend on the circumstances surrounding the presentation. This should not, however, keep the team from carefully considering and consciously deciding this and all of the other matters we've touched upon.

7.2.1.4 *Presentations are Visual Events*

Just as the team needs to know the audience, it should also try to know the setting in which they will be presenting. Some rooms will support certain types of visual aids, while others will not. At the earliest stages of the presentation planning, the design team should find out what devices (e.g., 35mm slide projectors, overhead projectors, computer connections) are available and the general setting of the room in which it will be presenting. This includes the size and capacity, lighting, seating, and other factors. Even if a particular device or setup is said to be available, it is always wise to bring along a backup, such as preparing transparencies to back up a slide presentation.

There are other tips and pointers to keep in mind about visual aids, including:

- Avoid using too many slides or graphics. A reasonable estimate of the number of slides that can be covered is 1–2 slides per minute of presentation. If too many slides are "planned," the presenter(s) will find themselves rushing through the slides in the hope of finishing. This makes for a far worse talk than a smaller selection wisely used.
- Beware of "clutter." Slides should be used to highlight key points; they are not a direct substitute for the reasoning of the final report. The speaker should be able to expand upon the points in the slides.
- Make points clearly, directly, and simply. Slides that are too flashy or clever tend to detract from a presentation.
- Use color skillfully. Current computer-based packages support many colors and fonts, but their defaults are often quite appropriate. Also, avoid weird or clashing colors in professional presentations.
- When describing the outcomes of the design process, do not reproduce the tools themselves (e.g., objectives tree, large morph charts). Rather, highlight selected points of the outcomes. This is a situation where it makes much more sense to refer the audience to the final report for more detailed information.

It is worth remembering that audiences tend to read visual aids as the speaker is talking. Therefore, the speaker does not need to read or quote those slides. The visual aids can be simpler (and more elegant) in their content because the visual aids reinforce the speaker, rather than the other way around.

Finally, if design drawings are being reproduced and shown, the size and distance of the audience must be considered carefully. Many line drawings are difficult to see and interpret in large presentations.

7.2.2 Design Reviews

A design review is a unique type of presentation that is quite different from all the others that a design team is likely to be called upon to do. It is also particularly challenging and useful to the design team. As such, a few points about design reviews are worth noting.

A design review is typically a long meeting at which the team presents its design choices in detail to an audience of technical professionals who are there to: assess the design, raise questions, and offer suggestions. In some cases, this is the best opportunity that the team will have to get the undivided attention of professionals about their design project. It is often also a scary and worrisome presentation for the design team because its members are being asked to answer pointed questions and defend their design. A design review thus offers both a challenge and an opportunity to the team, giving it a chance to display its technical knowledge and its skills in constructive conflict.

A design review is intended to be a full and frank exploration of the design, and it should expose the implications of solving the design problem at hand or even of creating new ones. A typical design review will consist of a briefing by the team on the nature of the problem being addressed, which is followed with an extensive presentation of the proposed solution. In the cases of artifacts, the team will often present an organized set of drawings or sketches that allows its audience to understand and question the team's design choices. In some cases, these materials may be provided to the attendees in advance.

As the meeting continues, all questions and technical issues should be fully explored in a positive, frank environment. This calls for the team to withhold the natural defensiveness that comes from having its design work questioned and challenged. In many cases the questions raised can be answered by the team, but in many others they cannot. Depending on the nature of the meeting, the team may call upon the expertise of all of the participants to suggest new ways to frame the problem or even the design itself.

Not surprisingly, such reviews can last several hours, or even an entire day. An important challenge for the team is to decide when a matter has been sufficiently covered and move on. This is a real challenge, since there is a natural temptation to move on when the issues raised suggest that the design must be changed in ways the team doesn't like. It is important to resist this urge and be sure that time management doesn't become a cover for hiding from criticism.

A final point about design reviews is the need to remember that conflict in the realm of ideas is generally constructive, while personality-oriented criticism tends to be destructive. Given the heat and light that sometimes arise at design reviews, team leaders and team members (as well as the members of the audience) must themselves act to continually refocus the meeting on the design, and not on the designers.

7.3   DESIGN DRAWINGS AND FABRICATION SPECIFICATIONS: OTHERS WILL BUILD WHAT YOU DESIGN

Beyond communicating with the client about the project, a design team must also communicate, even if only indirectly and through the client, with the maker or manufacturer of the artifact that has been designed. This is, in a very real sense, "where the design rubber meets the road" because someone whom the design team may never meet is going to build what the team has designed. Generally, the only instructions available to the maker are those representations or descriptions of the designed object that are included in the final report. This suggests that these representations and descriptions should be complete, unambiguous, clear, and readily understood. What can we do, as designers, to ensure that our product descriptions will result in our designed object being built just as we designed it?

The answer sounds deceptively simple: When communicating our design results to a manufacturer, we need to think long and hard about the kind of fabrication specifications (that we introduced in Chapter 1) we are writing. In addition, as part of the specification-writing process, we must take great care in rendering our final design drawings. As we will note below, this means paying particular attention to the various kinds of drawings that are done during a design project and to the different standards that are associated with final design drawings.

7.3.1   Design Drawings

We first turn to design drawings, which can include sketches, freehand drawings, and computer-aided-design-and-drafting (CADD) models that extend from simple wire-frame drawings (e.g., something very much like stick figures) through elaborate solid models (e.g., elaborate "paintings" that include color and three-dimensional perspective). Drawing is very important in design, especially in mechanical design, because a great deal of information is created and transmitted in the drawing process.

In historical terms, we are talking about the process of putting "marks on paper." These marks include both sketches and drawing and *marginalia*, that is, notes written in the margins. The sketches are of objects and their associated functions, as well as related plots and graphs. The marginalia include notes in text form, lists, dimensions, and calculations. Thus, the drawings enable a parallel display of information as they can be surrounded with adjacent notes, smaller pictures, formulas, and other pointers to ideas related to the object being drawn and designed. We see, here, that putting notes next to a sketch is a powerful way to organize information, certainly more powerful than the linear, sequential arrangement imposed by the structure of sentences and paragraphs. We show an example that illustrates some of these features in Figure 7.1. This is a sketch made by a designer working on the packaging, consisting of a plastic envelope and the electrical contacts, needed to accept the batteries providing power for a computer clock. The designer has written down some manufacturing notes adjacent to the drawing of the spring contact. Further, it would not be unusual for the designer to have scribbled modeling notes (e.g., "model the spring as a cantilever of stiffness . . ."), or calculations (e.g., calculating the spring stiffness from the cantilever beam model), or other information relating to the unfolding design. Note that some of this information also readily translates into aspects of fabrication specifications.

Marginalia of all sorts are familiar sights to anyone who has worked in an engineering environment. We often draw pictures and surround them with text and equations. We also draw sketches in the margins of documents, to elaborate a verbal description, to fortify understanding, to indicate more emphatically a coordinate system or sign convention. Thus, it should come as no surprise that sketches

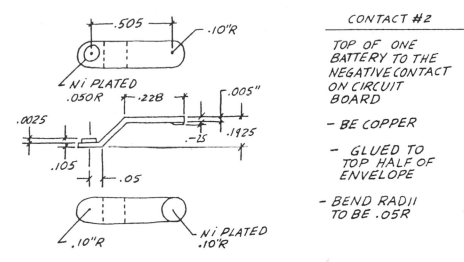

Figure 7.1 Design information adjacent to a sketch of the designed object (after Ullman, Wood, and Craig 1990).

and drawings are essential to engineering design. (It is interesting that while some classic engineering design textbooks stress the importance of graphical communication, drawing and graphics seem to have vanished from engineering curricula!) In some fields—for example, architecture—sketching, geometry, perspective, and visualization are acknowledged as the very underpinnings of the field.

Of particular importance to the designer is the fact that graphic images are used to communicate with the external environment, that is, with other designers, the client, and the manufacturing organization. Drawings are used in the design process in several different ways, including to:

- serve as a launching pad for a brand new design;
- support the analysis of a design as it evolves;
- simulate the behavior or performance of a design;
- provide a record of the shape or geometry of a design;
- facilitate the communication of design ideas among designers;
- ensure that a design is complete (as a drawing and its associated marginalia may remind us of still-undone parts of that design; and
- communicate the final design to the manufacturing specialists.

As a result of the many uses for sketches and drawings, there are several different kinds of drawings that can be formally identified in the design process. One list of the kinds of design drawings is strongly evocative of mechanical product design:

- *Layout drawings* are working drawings that show the major parts or components of a device and their relationship (see Figure 7.2). They are usually drawn to scale, do not show tolerances (see below), and are subject to change as the design process continues.

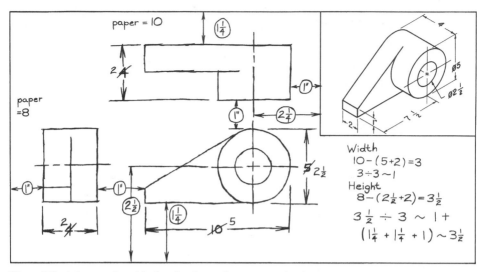

Figure 7.2 A *layout drawing* that has been drawn to scale, does not show tolerances, and is certainly subject to change as the design process continues. From (Boyer et al. 1991).

- *Detail drawings* show the individual parts or components of a device and their relationship (see Figure 7.3). These drawings must be toleranced, must indicate materials and any special processing requirements, are only changed when a formal "change order" provides authorization, and are drawn in conformance with existing drawing standards (which are also discussed below).

- *Assembly drawings* show how the individual parts or components of a device fit together. An *exploded view* is commonly used to show such "fit" relationships (see Figure 7.4). Components are identified by a part number or an entry on an attached *bill of materials*, and they may include detail drawings if the major views cannot show all of the required information.

In describing the three principal kinds of mechanical design drawings, we have used some technical terms that need definition. First, drawings show *tolerances* when they define the permissible range of variation in critical or sensitive dimensions. As a practical matter, it is literally impossible to make any two objects to be *exactly* the same. They may appear to be the same because of the limits of our ability to distinguish differences at extremely small or fine resolution. However, when we are producing many copies of the same thing, we want them to function pretty much the same way, so we must limit as best we can any variation from their ideally designed form. That's why we impose tolerances that prescribe limits on the manufacturer and what he produces.

We have also noted the existence of drawing standards. *Standards* explicitly articulate the best current engineering practices in routine or common design situations. Thus, standards indicate performance bars that must be met for drawings (e.g, ANSI Y14.5M–1994 *Dimensions and Tolerancing*), for the fire safety of buildings built within the United States (e.g., the *Life Safety Code* of the National Fire Protection Association), for boilers (e.g., the ASME *Pressure Vessel Code*),

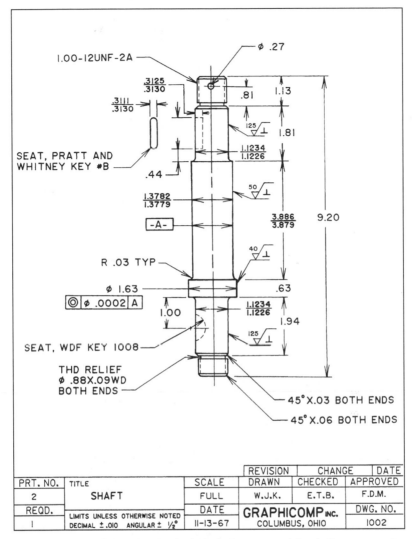

Figure 7.3 A *detail drawing* that includes tolerances and that indicates materials and lists special processing requirements. It was drawn in conformance with ANSI drawing standards. From (Boyer et al. 1991).

and so on. The American National Standards Institute (ANSI) serves as a clearinghouse for the individual standards written by professional societies (e.g., ASME, IEEE) and associations (e.g., NFPA, AISC) that govern various phases of design. ANSI also serves as the national spokesman for the United States in working with other countries and groups of countries (e.g., the European Union) to ensure compatibility and consistency wherever possible. A complete listing of U.S. product standards can be found in the *Product Standards Index*.

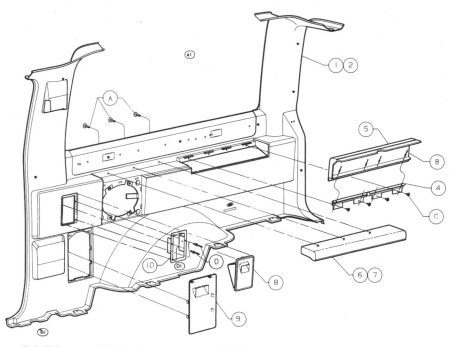

Figure 7.4 This *assembly drawing* uses an *exploded view* to show how some of the individual parts of an automobile fit together. Components are identified by a part number or an entry on an attached bill of materials (not shown here). From (Boyer et al. 1991)

7.3.2   Fabrication Specifications

As we noted in Chapter 1, the endpoint of a successful design project is the set of plans that form the basis on which the designed artifact will be built. It is not enough to say that this set of plans, which we have identified as the fabrication specifications, and which includes the final design drawings, must be clear, well organized, neat, and orderly. There are some very specific properties we want the fabrication specifications to have, namely, they should be *unambiguous* (i.e., the role and place of each and every component and part must be unmistakable); *complete* (i.e., comprehensive and entire in their scope); and *transparent* (i.e., readily understood by the manufacturer or fabricator).

We require fabrication specifications to have these characteristics because we want to make it possible for the designed artifact to be built by someone totally unconnected to the designer or the design process. Further, that artifact must perform just as the designer intended. Remember, this means that the designers are not there to catch errors or to make suggestions, and that the maker cannot turn around to seek clarification or ask on-the-spot questions. We are long since past the day when most designers were also craftsmen who made what they designed. As a result, we can no longer allow designers much latitude or short-hand in specifying their design work because they are unlikely to be involved in the actual manufacture of the design result.

Fabrication specifications are normally proposed and written in the detailed design stage (cf. Chart 2.4). Since our primary focus is conceptual design, we will not discuss fabrication specifcations in depth. However, there are some aspects that are worth anticipating even early in the design process. One is that many of the components and parts that will be specified are likely to be purchased from vendors, such as automobile springs, rocket O-rings, DRAM chips, and so on. This means that a great deal of detailed, disciplinary knowledge comes into play. This detailed knowledge is often critically important to the lives of a design and its users. For example, many well-known catastrophic failures have resulted from inappropriate parts being specified, including the Hyatt Regency walkway connections, the Challenger O-rings, and the roof bracing of the Hartford Coliseum. The devil really is in the details!

Of course, it sometimes is the manufacture or use of a device that exposes deficiencies that were not anticipated in the original design. That is, the way that designed objects are used and maintained produces results that were not foreseen. For example, consider failures such as the F–104 fighter plane that became known as "the widowmaker" because test pilots found that they could do flight maneuvers that the plane's designers did not anticipate (and that they also didn't think were appropriate, when they finally learned of these pilot maneuvers!). In 1979 an American Airlines DC–10 crashed because the plane's owners did a maintenance procedure in a way that completely undermined the design of the engine's supporting structures and their connections to the plane's wings. How much fortune-telling ability must a designer have? How far into the future, and how well, must a designer foresee the uses and misuses to which her work will be put? There are clearly ethical and legal issues here, but for the moment our intent is only to convey the fact that design details, such as fabrication specifications, are really important.

Given that many parts and components can be bought from catalogues, even while others are made anew, what sort of information must the designer include in a fabrication specification? Briefly, there are many kinds of requirements that can be specified in a fabrication specification, some of which are:

- the physical dimensions
- the kinds of materials to be used
- unusual assembly conditions (e.g, bridge construction scaffolding)
- operating conditions (in the anticipated use environment)
- operating parameters (defining the artifact's response and behavior)
- maintenance and life-cycle requirements
- reliability requirements
- packaging requirements
- shipping requirements
- external markings, especially usage and warning labels
- unusual or special needs (e.g., must use synthetic motor oil)

This relatively short list of the different kinds of issues that must be addressed in a fabrication specification really does make the point about the requirements we have for the properties of such a specification. The specification of the kind of spring action we see in a nail clipper might not seem a big deal, but the springs in the landing structure of a commercial aircraft had better be specified pretty darn carefully!

One final note here. In the same way that there are different ways to write design specifications (cf. Section 1.2 and Chapter 5), we can anticipate that there are different ways of writing the fabrication specifications. We might simply specify a particular part and its number in a vendor's catalogue, which would be a *prescriptive specification*; or we might specify a class of devices that do certain things, which would be a *procedural specification*; or we might simply leave it up to a supplier or the fabricator to insert something that achieves a certain function to a specified level, which would be a *performance specification*.

7.3.3   Philosophical Notes on Specifications and Drawings and Pictures

Inasmuch as there are so many standards that define practices in so many different engineering disciplines and domains, and since these are more likely to come into play in detailed design, rather than in conceptual design, we will close out our discussion of design drawings and fabrication specifications with a few philosophical notes.

The first is about the different approaches used by the different engineering disciplines. These differences arise in large part because of the different ways that each of the disciplines has grown and evolved, and they continue because of various needs that each discipline has. In the domain of mechanical design, for example, in order to make a complex piece that has a large number of components that fit together under extremely tight tolerances, there is no way to complete that design other than by constructing the sequence of drawings described above. There is no matching topological equivalent, that is, we cannot usually specify a mechanical device well enough to make it by drawing springs, masses, dampers (that are also called dashpots), pistons, etc. We have to draw explicit depictions of the actual devices. In the field of circuit design, on the other hand, both practice and technology have merged to the point that a circuit designer may be finished when she has drawn a circuit diagram, the analogy of the spring-mass-damper sketch. We can't discuss here the many complex reasons that practices differ so much among the engineering disciplines, or the various practices themselves. Nevertheless, it is important that designers be aware that while there are habits and styles of thought that are common to the design enterprise, there are practices and standards unique to each discipline, and it is the designer's responsibility to learn and use them wisely.

We also want to reinforce the theme that some external, pictorial representation, in whatever medium, is absolutely essential for the successful completion of all but the most trivial of designs. Think of how often we pick up a pencil or a piece of chalk to sketch something as we explain it, whether to other designers, students, teachers, and so on. Perhaps this occurs more often in mechanical or structural design because the corresponding artifacts quite often have forms and topologies that make their functions rather evident. Think, for example, of such mechanical devices as gears, levers, and pulleys. Think, too, of beams, columns, arches, and dams. This evocation of function through form is not always clear.

Sometimes we use more abstract drawings to show functional verisimilitude without the detail of sketches that are based on physical forms. Three examples of this sort of drawing abstraction reflect the kinds of discipline-dependent differences we discussed above: (1) the use of circuit diagrams to represent electronic devices; (2) the use of flowcharts to represent chemical-engineering-process plant designs, and (3) the use of block diagrams (and their corresponding algebras) to represent control systems. These pictures and charts and diagrams, with all of the different levels of abstraction we have seen, only serve to extend our limited abilities, as humans, to flesh out complicated pictures that exist solely within our minds.

Perhaps this is no more than a reflection of a more accurate translation of a favorite Chinese proverb, "One showing is worth a hundred sayings." It may also reflect a German proverb, "The eyes believe themselves; the ears believe other people." In fact, a good sketch or rendering can be very persuasive, especially when a design concept is new or controversial. Drawings serve as excellent means of grouping information because their nature allows us (on a pad, at the board, and soon in CADD programs) to put additional information about an object in an area adjacent to its "home" in a drawing. This can be done for the design of a complex object as a whole, or on a more localized, part-by-part basis. Again, drawings and diagrams are very effective at making geometrical and topological information very explicit. However, we do have to remember that drawings and pictures are limited in their ability to express the ordering of information, either in a chain of logic or in time.

Our final observation in this regard is that we have not made any reference to photographic images. Certainly, photos have much of the content and impact that we ascribed to other graphical descriptions, but they are not widely used in engineering design. One possible exception is the use of optical lithographic techniques to lay out very large scale integrated (VLSI) circuits, wherein a photography-like process is used. It is also the case that we are increasingly collecting data by photographic means (e.g., geographic data obtained from satellites). With computer-based scanning and enhancement techniques, we should expect that design information will be represented and used in this way. One sign of this trend is the increasing interest in geographic information systems (GIS), which are highly specialized database systems designed to manage and display information referenced to global geographic coordinates. It is easy to envision that satellite photos will be used together with GIS and other computer-based design tools in design projects involving large distances and spaces (e.g., hazardous-waste disposal sites and urban transportation systems). Thus, we should not forget photography as a form of graphical representation of design knowledge, alongside sketches, drawings, and diagrams.

7.4   FINAL REPORT ELEMENTS FOR THE XELA-AID DESIGN PROJECT

As required in most design projects, the student teams responsible for the chicken coop designs reported their results in the form of final reports and oral presentations. In this section we look briefly at some of the intermediate work products associated with their reports to gain further insight into some of the "do's and don'ts" discussed in Section 7.1. We will also report the conceptual design that was ultimately selected and built in San Martin Chiquito.

7.4.1   Rough outlines of two project reports

The two teams we have followed each prepared a rough outline as a first step in laying out the report structure. Table 7.1 shows the rough outline of one of the teams, and Table 7.2 shows that for the other team.

These two outlines display both similarities and differences. The first team, for example, has dedicated several sections to justifying their final design, while the second team has organized itself around process. Both teams have relegated most of their sketches and drawings to appendices, although the second team has put building instructions in the body of the report. This reflects the freedom that teams have in deciding on an appropriate structure to convey their design results. This freedom, however, does not excuse them from having a logical ordering that allows the reader to understand the nature of the problem or the benefits of their solution.

Table 7.1 A rough outline for one of the chicken coop teams. The rough outline should show the overall structure of the report in a way that allows team members to divide up work with little or no unintended duplication. The structure should also proceed in a clear and logical manner. Does it for this report?

I.	Introduction
II.	Description of what needed to be accomplished
	A. weighted objectives tree
	B. weighted objectives tree justifications
III.	Generation and evaluation of alternatives
	A. morphological chart
	B. metrics chart
IV.	Design results
V.	Justification of selection of coop attributes: Evaluation of our solution
VI.	Project management
	A. work breakdown structure
	B. schedule
	C. budgeting
VII.	Conclusions
	A. Results of our analysis of the problem
	B. Insights for next time
	C. Chicken coop suggestions
VIII.	References and endnotes
Appendix:	Price analysis in American dollars
Appendix:	Fabrication specifications

Table 7.2. A rough outline for another of the chicken coop teams. This rough outline also shows the overall structure—the report clearly is focused on reporting the process by which the design was arrived at as well as the actual outcome. This puts the team at some risk of writing a "history of the project" unless great care is exercised in writing the draft. This can be handled easily in the topic sentence outline.

I.	Introduction	
II.	Client statement	
III.	Design process:	
	A.	objectives tree
	B.	weighted objectives tree
	C.	functional analysis
	D.	morphological chart
	E.	metrics and testing
IV.	Final design selection	
V.	Building	
VI.	References	
Appendix:	Research on poultry housing	
Appendix:	Worksheet used to rank objectives	
Appendix:	Final design diagrams	

A second point to make regarding the structure of these final reports is to note just how much of them could have been done during the course of the project. Each of the reports uses the formal design tools discussed in the previous chapters to document its decision process. Because of this, the teams could—and should—have been tracking and organizing their outcomes in order to facilitate the writing of the final report.

A final point about outlines is that neither is sufficiently adequate to translate directly into an actual report. There are many issues that could be considered in more than one section, and others that may not be covered at all. Unless the team continues with a topic sentence outline or some other detailed plan, the first draft of the final report will require an unnecessarily high degree of editing.

7.4.2   Expanding the Outline for the Xela-Aid Project

Table 7.3 shows an excerpt from the topic sentence outline prepared by one of the student design teams. The overall outline was eight pages long, single spaced. Notice that while each entry is not in itself a complete sentence, it is very easy to see what the specific point of that entry would be. At this level of detail, it is relatively straightforward to ascertain which points are either redundant or inadequately covered.

The TSO enables the team to see not only what will be covered within each section, but also within each paragraph of the report. It also permits team members to take issue with or make suggestions about a section before someone has invested the time to actually write and "wordsmith" it. For example, the team's definitions of

Table 7.3 An excerpt from a student team's TSO showing the general approach for moving from the rough outline toward a coherent first draft. Note a tendency to write in terms of chronology (e.g., "Now that we had…") and some loose definitions. The team could ask whether or not one paragraph is sufficient for each idea being carried. For example, there is no specific paragraph that identifies the particular morphological chart the team used.

V.   Explanation and evaluation of feasible alternatives we considered
  A.   To assist us in the determination of feasible alternatives we created what is known as a morphological chart
    1.   A paragraph describing a morph chart and what it consists of
      a)   Functions
      b)   Solutions to functions (means)
  B.   Once our chart was complete we began generating alternatives based on it
    1.   A paragraph describing use of the morph chart
      a)   Ruling out possibilities based on inconsistencies
      b)   Starting with the functions with the most limiting solutions
      c)   The production of valid design alternatives from combinations of solutions
  C.   Now that we had a set of valid alternatives we developed methods for determining to what degree each alternative could meet our objectives
    1.   A paragraph describing the definition of a metric
      a)   A methodology designed to allow us to qualitatively or quantitatively measure an aspect of a design
      b)   Each metric is designed to measure one aspect of the design
      c)   Metrics are often derived from objectives
        i)   Often used to determine the level at which an objective is being met
      d)   Data collection is used to make determinations about which design components are the best (meet the most objectives) and thus determine the final design
  D.   The information we collected on chickens, coops, predators, climate, and the village conditions also influenced our design decisions.
    1.   A paragraph or two describing the key points of our research . . .

---

metrics are somewhat unclear, and could be challenged by a nit-picking reader (such as a professor or a technical manager). It is also not at all clear that all the ideas being conveyed in some of the paragraphs couldn't be better covered by separating them into two paragraphs. For example, the definition of a morph chart, as opposed to the display of the team's own morph chart, might be better given separately.

Notice also that the team has adopted a historical approach to the process, which is very much at odds with recommended style. For example, "once our chart was complete we began . . ." is a red flag that the team is documenting the passing of time and events, not the design process.

Fortunately, because the team has invested effort in the TSO, it is relatively easy to make changes. If the team were to decide, for example, to move the information on chickens, coops, etc., to an appendix, this material can be easily included and used by reference to that section.

7.4.3   The Final Outcome: A Chicken Coop in San Martin Chiquito

Having used the chicken coop in San Martin Chiquito as a design example, and particularly as one in which there were often either shortcomings or areas for improvement, it seems only fair and appropriate to describe the final outcome. Both teams offered interesting designs, one a "chicken factory" conceptual design, the other based on a fenced-in area with smaller, less expensive nesting and perching areas. After consulting with the client on a number of factors, including cost, ease of construction, and productivity, the fenced-in concept was adopted. During the summer of 1997, following the completion of the design, a team of four students and their faculty advisor (who happens to be one of the authors) traveled to Guatemala as part of a Xela-Aid team and built the coop, along with a greenhouse and a weaving building.

Figures 7.5 and 7.6 show photographs of the coop both under construction and very near to poultry occupancy. A number of changes and refinements from the initial design can be seen in the photographs, which is not uncommon in the transition from a conceptual design to its final implementation. The original design, for example, called for a cinder-block coop building. As a practical matter, it turned out to be both cheaper and easier to build the coop using wood and corrugated metal as the primary construction materials. The conceptual design did not adequately address the issue of preventing the wood in the base of the coop, which was untreated, from rotting in the ground. However, on-site, a system of concrete pads was developed, onto which the wood was bolted. A further change was that angle iron was used for the fence posts because it was available

Figure 7.5 This photograph shows the chicken coop under construction in San Martin Chiquito. Notice the nest boxes on the left, and the perches on the right side. The women in the foreground include Amalia Vazquez, the leader of the women's cooperative, i.e., the client.

Figure 7.6 This photograph shows the chicken coop during the final stages immediately prior to occupancy by the poultry. The gate still needed attachment to the fence, and some cosmetic changes had not yet been made. In the background some of the construction materials for a weaving cooperative can be seen. The dogs in the foreground were removed and replaced with 30 chickens, tripling poultry and egg production

in Guatemala and considerably easier to use than the triple rebar construction alluded to in our previous discussions of this design project.

At first glance, these and other changes made the final coop appear to be very different than what was originally proposed by the students. However, the basic concept did not change. A look at Figure 7.5 very clearly shows both the trays that make up the nesting boxes and even the perching posts.

Today the women's cooperative in San Martin Chiquito boasts a constant production of 30 chickens and the associated egg production, a tremendous improvement as a consequence of a student design project.

7.5 MANAGING THE PROJECT ENDGAME

In this section we present some final remarks on managing the documentation activities and on closing out the project. Of particular importance to teams that want to use their experience on one project as a basis for improving their performance on the next project is a discussion of the post-project audit.

7.5.1 Team Writing is a Dynamic Event

Most of us have considerable experience in writing papers by ourselves, all alone. This may include term papers for school, technical memoranda, lab write-ups, and even creative writing or journalistic experiences. Documenting a design in a team setting is, however, a fundamentally different activity than writing a paper alone. These differences turn on our dependence on co-authors, the technical demands, and the need to ensure a uniform style.

When writing as a team, we can only be sure what others are writing if all of our writing assignments and their associated content are made explicit. This means that even if any one team member's personal writing style does not include detailed outlines and multiple drafts, all members should do outlines and rough drafts because they are essential for the team's success. This also makes the issues of responsibility and cooperation important in a way that is unique for most writers. Recall that in Chapter 4 we introduced the linear responsibility chart (LRC) as a means of ensuring the equitable and productive allocation of work. As part of the documentation phase, the LRC should be revised and updated.

Even with a fair allocation of work, the ultimate quality of the final report, an oral presentation, and other forms of documentation will reflect on the team as a whole and on each of its members. It is important, therefore, to allow sufficient time for each of the team members to read report drafts carefully. Equally important, the team must create an atmosphere in which the comments and suggestions of others are treated with respect and consideration. No one on the team should be exempt from reading the final report drafts, and no one's opinion should be treated with disrespect. Given the pressure under which final deliverables are often prepared, the atmosphere is, in some sense, a test of the group's culture and attitudes. The interpersonal dynamics should be monitored closely and managed carefully.

As we noted in Section 7.1, the team must also agree to a single voice. This is often quite difficult for teams to work toward, especially when several days' work is rewritten after so much work has been done. It is hard for any writer, and still harder for those who consider themselves skilled writers, to sublimate their styles and their egos to satisfy a team and its designated editor. Once again, all of the members of a team have to remember the team's overall goals.

Similarly, oral presentations demand that a team divide up the work fairly. Each team member must also recognize that other members may be presenting the team's work. In many cases, the presenter of a particular piece of the project may have had little to do with the element of the work being presented, or may even have opposed that approach. Once again, then, a central issue here is the need for mutual respect and appropriate action by the entire team.

7.5.2 Project Post-Audits: Next Time We Will . . .

In actual practice, most projects end not with the delivery to the client, but with a *project post-audit*: an organized review of the project, including the technical work, the management practices, the work load and assignments, and the final outcomes. This is an excellent practice to develop, even for student projects or activities where the team is disbanding completely. There is an old Kentucky saying to the effect that the second kick of the horse has no real educational value. The post-project audit is an opportunity to understand the horse better and learn where to stand next time.

The key issue in post-project audit is focus on doing an even better job next time around. As a practical matter, the post-project audit may be as simple as a meeting that takes one or two hours, or it may be part of a larger formal process that is directed by the design team's parent organization. Regardless of the scope or formal mechanism, if any, the basic post-audit process is simple:

- review the project goals;
- review the project processes, especially in terms of ordering of events;
- review the project plans, budgets, and use of resources; and
- review the outcomes.

Reviewing the project goals is particularly important for design projects, since design is a goal-oriented activity. If the project was supposed to solve problem A, then even an idea that results in earning a patent for solving problem B may not be viewed as a success. We can only evaluate a project in terms of what it set out to do. To this end, many of the problem definition tools and techniques should be reviewed as part of the post-audit.

Closely tied to reviewing the results of using the design and management tools is the useful notion of having the team consider the effectiveness of the tools themselves. Just as a toolbox may contain many items that are only useful some of the time, many of the formal methods and techniques that are presented in this book and elsewhere will be more effective in some situations than in others. No catalog of successes or failures by the authors will have the same purchase with the team as their own experience. Reflecting on what worked and what didn't, coming to grips with why a tool did or did not work, are both important elements of the post-project audit.

Analogously, reviewing the manner in which a team managed and controlled its work activities is also important to avoiding that "second kick of the horse." Most teams learn only how to organize activities, determine their sequence, assign the work, and monitor progress by experience and practice. Such experiences are much more valuable if they are reviewed and reconsidered after the fact. As with the design tools, not all of the management tools are equally useful in every setting (although some, such as the work breakdown structure, appear to useful in almost every situation). In commercial settings, reviewing both budgets and work assignments is critically important for planning future actions.

The final step is to review the outcome of the project, in terms of the goals and the processes used. While it is certainly useful to know whether or not the goals were achieved, it is important for the team members to ascertain whether this is a consequence of excessive resources, good planning and execution, or simply good luck. In the long run, only teams that learn good planning and execution are likely to have repeated successes.

Our final note is that the post-project audit is not, in and of itself, a tool for assigning blame or for pointing fingers. Many project and institutional settings have formal mechanisms for peer review and supervisory evaluation of team members, and they can be valuable means for highlighting individual strengths, weaknesses, and contributions. They can also provide team members with important insights that they can use to improve their work in design teams. However, individual performance reviews are not a central, or even a desirable aspect of the post-project audit. The audit is intended to show what the team and organization did right to make the project successful, or what must be done differently if the project was not successful.

7.6 NOTES

Section 7.3: The list of the kinds of fabrication specifications has been adapted from (Ertas and Jones 1993). Some of the failures mentioned, and many more, are detailed in (Schlager 1994). Much of the discussion of drawing is drawn from (Ullman, Wood, and Craig 1990) and (Dym 1994). The listing of kinds of design drawings is adapted from (Ullman 1997). The Chinese and German proverbs are from (Woodson 1966).

Section 7.4: The final results of the chicken coop design projects are drawn from (Connor et al. 1997) and (Gutierrez et al. 1997).

7.7 EXERCISES

7.1 How can a design team determine the audience for an oral presentation?

7.2 How does the composition of an audience affect the structure and content of a design team's presentation to that audience?

7.3 Determine the appropriate standards (e.g., ANSI standards) for presenting the design of the portable electric guitar of Exercise 3.2.

7.4 Determine whether there are any comparable appropriate standards for presenting the design of the rain forest project of Exercise 3.5.

7.5 In your role as the team leader of an HMCI engineering design team, you encounter the following situation while preparing the team's final report. Whenever Ken submits written materials documenting his work, David criticizes the writing so severely and personally that the other members of the team become quite uncomfortable. You recognize that Ken's work product does not meet your standards, but you also consider David's approach to be counterproductive. How might this conflict be resolved constructively?

7.6 You are a member of a design team that has, at last, completed its work on a design project. You have been asked by the team leader to organize a post-audit review of the team's work. Explain your strategy for conducting such an audit.

Chapter 8

Design for . . .

What are some of the consequences of our technical choices?

A central theme of this book is that engineering design is often done by teams rather than individuals. This idea reflects the recent experience of engineers in industrial settings throughout the world. Design teams usually include not only engineers, but also manufacturing experts (who may be industrial engineers), marketing and sales professionals, reliability experts, cost accountants, lawyers, and so on. Such teams are concerned with understanding and optimizing the product under development for its *entire life*, including its design, development, manufacturing, marketing, distribution, use, and, eventually, disposal. Concern with all of these areas and with their impact on the design process has come to be known as *concurrent engineering*. While it is beyond the scope of many smaller projects (e.g., the student design examples described in earlier chapters) to apply concurrent engineering techniques, it is important that engineering designers be aware of concurrent engineering and its implications for their designs.

One of the most important aspects of engineering in the modern commercial setting is the realization that the audience for a good design must include those who will build and maintain the designed artifact. Another important concept is that designs must usually meet economic or cost-related targets. These concepts are part of a more general notion, one that reflects the fact that engineers have always sought to realize various desirable attributes in their designs to some degree. This is often referred to as "design for X," where X is some attribute such as manufacturing, maintainability, reliability, or affordability. (Designers and engineers also refer to them with a different name, the *-ilities*, because many of these desirable attributes are expressed as nouns that have an "-ility" suffix.) We now focus on some of these "-ilities" or X's in the hope of raising the consciousness of engineering design teams about subjects that we do not treat here in great depth (e.g., disposal of used batteries for one product, or design for recyclability for another).

In keeping with the notion of the life cycle of a product described above, we will look first at design for manufacturing and assembly, then design for affordability, and finally design for reliability and maintainability.

8.1   DESIGN FOR MANUFACTURING AND ASSEMBLY: CAN THIS THING BE MADE?

In many cases, the artifact that an engineering team designs will be produced or manufactured in quantity. In recent years, companies have come to learn that the design of a product can have an enormous impact on the cost of producing that artifact, on its resulting quality, and on its other characteristics. Toward this end, the notion of considering how a product is manufactured as part of the design process has come to be a central idea for globally competitive industries such as the automotive and consumer electronics industries. A significant part of this concern is based on the large number of products being manufactured, which allows for economies of scale, which we will discuss in Section 8.2. There is also a realization that for many products, the time it takes to get a product to the consumer, known as the *time to market*, defines a company's ability to shape the market. Design processes that incorporate manufacturing issues can be key elements in speeding products through to commercial production.

8.1.1   Design for Manufacturing (DFM)

Design for manufacturing (DFM) is design based on minimizing the costs of production and/or the time to market for a product, while maintaining an appropriate level of quality. The importance of maintaining an appropriate level of quality cannot be overstated because without an assurance of quality, DFM is reduced to simply producing the lowest-cost product.

DFM almost inevitably begins with the the formation of the design team. In commercial settings, design teams committed to DFM tend to be multidisciplinary, including not only engineers, but manufacturing managers, logistics specialists, cost accountants, and marketing and sales professionals. Each brings particular interests and experience to a design project, but all must focus on the project itself and move beyond their primary expertise. In many world-class companies, such multidisciplinary teams have become the de facto standard of the modern design organization.

Manufacturing and design tend to have iterative interactions during product development. That is, the design team will learn either of a problem in producing a proposed design or of an opportunity to reduce design cost or timing, as a result of which the team will reconsider its design. Similarly, the design team may be able to suggest alternative production approaches that will lead manufacturing specialists to consider restructuring processes. In order to achieve fruitful and synergistic interaction between the manufacturing and design processes, it is important that DFM be considered in each and every one of the design phases, including the early conceptual design stages.

One basic methodology for DFM consists of six steps:

1. estimate the manufacturing costs for a given design alternative;
2. reduce the costs of components;
3. reduce the costs of assembly;
4. reduce the costs of supporting production;
5. consider the effects of DFM on other objectives; and
6. if the results are not acceptable, revise the design or the modifications once again.

This approach clearly depends upon an understanding of all the objectives of the design, else the iteration called for in Step 6 cannot meaningfully occur. An understanding of the economics of production is also required; some of the economic issues are discussed in Section 8.2. In addition to these, however, there are engineering and process decisions that can directly influence the cost of producing a product. Certain processes for shaping and forming metal, for example, can have dramatically higher costs than others and may only be called for when particular engineering needs must be met. Similarly, some types of electronic circuits can be made with high-volume, high-speed production machines, while others require hand assembly. Some design choices that would require higher costs for small production runs may actually be less expensive if the design can also be used for another, higher-volume purpose. In each of these cases, it is only the combination of deep knowledge about manufacturing techniques with a wealth of design experience that can be counted on to successfully complete a design.

8.1.2   Design for Assembly (DFA)

A related, but formally different brand of design for X is *design for assembly*. Assembly refers to the ease with which the various parts, components, and subsystems are joined, attached, or otherwise grouped together to form the final product. Assembly can be characterized as consisting of a set of processes by which the assembler (1) handles parts or components (i.e., retrieves and positions them appropriately relative to each other), and (2) inserts (or mates or combines) the parts into a finished subsystem or system. For example, assembling a ball point pen might require that the ink-holding unit be inserted into the tube that forms the handgrip, and that caps be attached to each end. This assembly process can be done in a number of ways, and the designer needs to consider approaches that will make it possible for the manufacturer to reduce the costs of assembly while maintaining high quality in the finished product. Clearly then, assembly is a key aspect of manufacturing and must be considered either as part of design for manufacturing or as a separate design task.

Because of its central place in manufacturing, a great deal of thought has been put into development of guidelines and techniques for making assembly more effective and efficient. Some of the matters that are typically considered are:

1. *Limiting the number of components to the fewest that are essential to the working of the finished product.* Among other things, this implies that the designer will differentiate between parts that could be eliminated by combining other parts and those that must be distinct as a matter of necessity. The usual rules for this are to note:
   - which parts must move relative to one another;
   - which parts must be made of different materials (for reasons of strength, for example, or insulation); and
   - which parts must be separated in order to allow for assembly to proceed.

2. *Using standard fasteners and integrating fasteners into the product itself.* By reducing the number and type of fasteners, the designer allows the assembler to construct a product without having to retrieve as many components and parts. Using standard fasteners also allows the assembler to develop standard routines for component assembly, including mechanization. The designer should also consider that fasteners tend to concentrate stresses in themselves and so may cause reliability concerns of the sort we will discuss in Section 8.3.

3. *Designing the product to have a base component on which other components can be located*, and designing for the assembly to proceed with as little motion of the base component as possible. This guideline enables an assembler (whether human or machine) to work to a fixed reference point in the assembly process and to minimize the degree to which the assembler must reset reference points.

4. *Designing the product to have components that facilitate retrieval and assembly.* This may include elements of detailed design that, for example, reduce the tendency of parts and subassemblies to become tangled with one another, or designing parts that are symmetric, so that once retrieved they can be assembled without turning to a preferred end or orientation.

5. *Designing the product and its component parts to maximize accessibility, during both manufacturing and subsequent repairs and maintenance.* While it is important that the components be efficient in their use of space, the designer must balance this need with the ability of the assembler or mechanic to gain access to and manipulate parts, for both initial fabrication and later replacement.

While these guidelines and heuristics represent only a small set of the design considerations that make up design for assembly, they do provide a starting point for thinking about both DFA and DFM.

8.1.3   The Bill of Materials (BOM)

Effective design for manufacturing also requires a deep understanding of production processes and technologies that may not be currently used by the client, including the use of robotics and other advanced technologies. Among the most important of these technologies are the ways that manufacturers plan and control their inventories. One of the most common techniques used is a process called Materials Requirements Planning (MRP). This technique utilizes the assembly drawings discussed in Chapter 7 to develop a Bill of Materials (BOM) and an assembly chart, known as a "gozinto" chart, which shows the order in which the parts on the BOM are put together. The BOM is a list of all the parts, including the quantity of each part required to assemble a designed object. We might think of the BOM as being a kind of recipe that specifies (1) all of the ingredients that are needed, (2) the precise quantities needed to make a specified lot size, and, when paired with the "gozinto" chart, (3) the process for putting the ingredients together.

When a company has determined the size and timing of their production schedule, the BOM is used to determine the size and timing of inventory orders. (Most companies now use *just in time* delivery of parts as they seek to avoid carrying large inventories of parts because these are paid for but not generating revenues until after they are assembled and shipped.) The importance of the assembly drawings and the BOM in managing the production process cannot be overstated. To be effective, the design team must not only develop accurate methods of reporting their design, but the entire organization must be committed to the discipline that any engineering design changes, or *engineering change orders*, will be reported accurately and thoroughly to *all* the affected parties. In Section 8.2, we will see that the BOM is also useful in estimating some of the costs of producing the designed artifact.

A final point to note is that manufacturing concerns include both logistics and distribution, so that these elements have also become an important part of design for manufacturing. One of the major changes in how business is done today is that companies now work toward forging links between the suppliers of materials needed to make a product, the fabricators who manufacture that product, and the channels needed to efficiently distribute the finished product. This set of related activities, often referred to as the *supply chain*, requires a designer to understand elements of the entire product cycle. It is beyond our scope to explore the role of supply chain management in design, except to note that in many industries, successful designers understand not only their own production and manufacturing processes, but also those of their suppliers and their customers. This requirement for integrated understanding of commercial processes will surely increase in the future.

8.2   DESIGN FOR AFFORDABILITY: HOW MUCH DOES THIS THING COST?

To be able to afford something is, according to the dictionary, to be able to bear the cost of something or to pay the price for it. In the design context, whether or not we can afford something is an issue that will be faced by the client (e.g., can I afford to make this product?, or perhaps, can I afford not to make it?), the manufacturer (e.g., can I afford to make this at a given price?), and the user (e.g., can I afford to buy this product?). Thus, affordability is really about expressing an important dimension of the object, artifact, or system being designed in terms that all of the stakeholders can recognize and understand, that is, in terms of money. This dimension is typically covered in the field known as *engineering economics*. Engineering and economics have been closely tied for almost as long as the two fields have each existed. Indeed, economists recognize that engineers were the developers of a number of important elements of economic theory. For example, what economists call utility theory and price discrimination were both first articulated by the 19th-century engineer Jules Dupuit, and location theory was developed by a civil engineer named Arthur M. Wellington. In fact, Wellington is credited with the definition of engineering as "the art of doing that well with one dollar which any bungler can do with two." This linking of engineering and economics should come as no surprise to the designer, since it is a rare project for which money really is "no object."

Engineering economics is concerned with understanding the economic or financial implications of engineering decisions, including choosing among alternatives (e.g., cost-benefit analysis), deciding when and if to replace machines or other systems (replacement analysis), and predicting the full costs of devices over the period of time that they will be owned and used (life-cycle analysis). These topics can easily fill entire courses in an engineering curriculum and are well beyond the scope of this chapter's discussions. However, there are some topics that are so important to the designer that they must be introduced, even if only briefly. Perhaps the most important of these topics is that known as the time value of money. A second topic of great importance is cost estimation. Without at least a rudimentary knowledge of these, design and engineering teams are likely to make good choices about designs only by luck.

8.2.1   The Time Value of Money

If someone were to offer us $100 today or that same $100 one year from now, we would almost certainly prefer to take the money *now*. Having the money sooner offers a number of advantages, including providing us with the ability to invest or otherwise use the money during the intervening year, eliminating the risk that the money might not be available next year, and eliminating the risk that inflation could reduce the purchasing power of the money during the intervening year. This simple example highlights one of the most important concepts in engineering economics, namely, the *time value of money*: money obtained sooner is more valuable than money obtained later, and money spent sooner is more costly than money spent later.

As indicated, the time value of money captures the effects of both foregone opportunities, referred to as *opportunity costs*, and *risk*. An opportunity cost is a measure of how much the deferred money could have earned in the intervening time. The risk captures both the risk that the money will be worth less (because of inflation) and the risk that the money will simply have become unavailable during the intervening time. Economists and financial professionals bundle together the extent of these risks and their associated lost opportunities in the *discount rate*. A discount rate acts much like an interest rate on a savings account or credit card, except that it views the money as being worth less to us today because of the risks and missed opportunities. The interest rate on a savings account measures how much a bank is willing to pay for the privilege of using our money in the coming year. An interest rate on a credit card measures how much we must pay the card issuer for the privilege of using its money; these rates vary, with higher charges assigned to customers with poorer credit ratings or with limited credit histories. Interest calculations thus typically show dollar amounts increasing from a given time onward or forward. Discount rates typically work in reverse, showing the value today of money that will be available at some point in the future.

Measuring risks and opportunity costs can be a complex process (and one well beyond our current scope), but as engineers we should remember the design decisions and choices we make today will translate into streams of prospective "financial events" that will occur at different times in the future. Some of these financial events are costs we will incur (e.g., for manufacturing or distribution), and some are benefits (e.g., revenues from sales). The more immediate those costs and benefits are, the more impact they will have on decisions made by both clients and users, and hence on us as designers.

If "the same money" has different values at different times, how do we make economic decisions in a rational and consistent way? The answer is that, given a discount rate and a set of future financial events or cash flows, both to us and away from us, we translate all of our events into a common time frame, either a current time or a future one. Consider again our choice between getting $100 today or next year. If the annual discount rate is 10%, then we would expect to need $110 dollars in a year in order compensate us for not having $100 today (or to purchase then what we could buy today with $100). That is, we would be getting the same amount of money if we say we'll accept $100 now or $110 a year from now.

We can also work this the other way, asking how much we need today to have the equivalent of $100 a year from now. Next year's $100 is worth the same amount as some amount of money, X, plus the 10% that X could earn in the coming year. That is, $1.10 \cdot \$X = \$100$. Solving for X shows that taking the $100 in a year is equivalent to getting and investing about $91 today. We won't go through all the various formulas and arithmetic associated with the time value of money, except to point out that we can carry out discounting calculations as far as we would like into the future. The principle would be the same. Thus, $100 promised for two years from now would be worth even less than $100 were today or even $100 promised for next year, since it wouldn't have accounted for being able to use the $100 for both years, or our being able to use the $10 earned in the first year.

Economists have developed standard approaches to discounting money and to determining the present value of future dollars, whether costs or benefits, and vice versa. Application of these formulas can become quite involved when factors such as inflation or unusual timing are involved, but virtually all such analysis is based around the relationship

$$PV = FV \left(\frac{1}{1+r} \right)^t \qquad (8.1)$$

where PV is the present value of the costs or benefits, FV is the future value, r is the discount rate, and t is the time period over which a cost is incurred or a benefit is realized. Consider again our decision concerning the worth of $100 next year. In this case, the future value, FV, is $100, the time period, t, is 1 year, and the discount rate, r, is 10% per year, or 0.10. If we substitute these values into Eqn. (8.1), the present value, PV, will be the $91 found above. In other words, an offer of $100 a year from now would be equivalent to an offer of $9 less today. The ability to translate future costs into present equivalent values, known as *discounting*, is very important in some design projects and can affect how we might choose among designs. We turn to that topic now.

8.2.2   The Time Value of Money Affects Design Choices

Imagine for a moment that you are asked to choose between two alternative vehicle designs for a transit agency. Design alternative A has a significantly higher initial purchase price, but design alternative B has higher operating costs over the life of the vehicle. In this case we must find a means of reconciling the different costs at different points in the lifetimes of the two choices. An obvious question is, how big are those cost differences and when do they occur? If design B can be bought for much less money than design A, its higher operating costs may not be as important because they are incurred so much later than the immediate initial saving. If, on the other hand, the operating costs of B are both much higher and occur relatively early in the vehicle's life, then the immediate savings gained by buying design B may be illusory. Making a rational choice in such a case requires that we understand how to properly analyze the time value of money (cf. Section 8.1.1) because we want to compare all of the costs in equivalent dollar values.

Now, imagine further that our designs not only have different purchase prices and differing operating and maintenance costs, but also have different expected lifetimes. This means that we must find a way to adjust all of the costs in some way that makes the values equivalent over the same time frame. Engineering economists have developed a methodology for doing that, although we will not provide a detailed description here. Called *equivalent uniform annual costs* (EUAC), it essentially treats all of the alternatives as though they are replaced with a one-for-one swap whenever they wear out. EUAC then transforms the resulting infinite series of replacements into a series of annual payments. The point to be drawn here is simply that the series of future costs and benefits of all design alternatives must be considered over the lifetime of each of the alternatives and then translated into a format that allows us to fairly compare those alternatives. The essential wisdom is that it is insufficient to look only at the initial purchase costs of design alternatives as a way of finding out what designs really cost. A true cost analysis requires that we consider the entire life cycle of a design.

8.2.3 Estimating Costs

In the previous section, we took it for granted that the cost of the final design is known, as are the operating and maintenance costs over the life of the device. In practice this is not usually a simple matter. It requires skills and experience that are beyond many novice designers (and students!). While the subject of cost estimating can easily consume an entire text, several points are relevant to designers at the conceptual design stage.

It is easy to say that the costs of a design typically include labor, materials, overhead, and profits for various stakeholders. However, this simple statement masks the complexity of detailing or structuring the cost of all but the simplest of artifacts. In many cases, estimating the cost of producing and distributing a design is an extremely difficult task. Here we limit ourselves to simply describing the principal elements that make up the cost categories listed just above.

Labor costs include payments to the employees who construct the artifact, as well as support personnel who perform necessary but seemingly invisible tasks such as answering the phone, filling orders, packing and shipping the product, etc. Labor costs also include a variety of *indirect costs* that are not immediately visible because they do not show up as payments made directly to employees. Such indirect costs are often termed *fringe benefits* because they are typically payments made to third parties on behalf of the employees. The fringe benefits include health and life insurance, retirement benefits, employers' contributions to Social Security, and other mandated payroll taxes. These indirect costs of labor are often neglected or overlooked by designers who are eager to estimate the cost of a design. In spite of their hidden nature, however, such indirect costs are as much as 50% of direct labor payments (i.e., wages) for many companies.

In Section 8.1 we discussed the importance of the Bill of Materials (BOM) in controlling inventories and managing the manufacturing of items. The BOM is also useful in estimating some of the costs associated with various designs, especially material costs. *Materials* include those directly used in building the device, but also may include intermediate materials and inventories that are used in ways that may not be obvious. For example, some inventory is wasted during manufacturing, while other inventory may be classified as being part of work-in-progress. The BOM provides a guideline for the number and type of parts that make up the device or object. It is particularly useful since it is developed directly from the assembly drawings described in Chapter 7, and so the BOM will reflect the designer's final intent.

We do need to exercise some care when we use a BOM to estimate costs since both labor and materials are subject to *economies of scale*, the notion that the production cost per (single) item, the *unit cost*, can often be reduced by making many identical copies, rather than simply hand-making "originals." The genius of Henry Ford's assembly line, wherein he came up with a way to make millions of copies of his cars, is a reflection of the economies of scale. Ford was able to lower his unit costs and sell all of the cars that he did because many more people could afford to buy them. Of course, Ford could only realize such economies of scale by developing new technological ideas, but that's still another story about how engineering and economics interact.

Overhead is the term applied to those costs incurred by a manufacturer that cannot be directly assigned to a single product. If, for example, a device is made in a factory that also produces twenty other products, the cost of the building, the machines, the janitorial staff, the electricity, etc., must somehow be shared or distributed among all of the 21 items. If each product was priced to ignore these overhead costs, the company would soon find itself unable to pay either for the building or for the services necessary to maintain it. Other elements of overhead include the salaries of executives, who are presumably using some share of their time to supervise each of the company's activities, and personnel and related costs of needed business functions such as accounting, billing, and advertising. While there are accounting standards that define cost categories and their attributes, precise estimates of overhead costs vary greatly with the structure and practices of the company in question. One company may have only a small number of products and a very lean organization, so that most costs can be directly attributed to the products made and sold, with only a small percentage remaining to be allocated as overhead. In other organizations, the overhead can be an amount equal to or greater than the labor costs that are directly assignable to one or more products. In many universities and colleges, for example, the overhead rate associated with research (which pays for laboratories, support staff, college presidents, and other essentials) runs as high as 65% of the researchers' salaries and benefits. The key point is that estimating the costs of producing a design requires careful consultation with clients or their suppliers.

Cost estimates produced during the conceptual stage of a design project are often quite inaccurate when compared with those made for detailed designs. In heavy construction projects, for example, an accuracy of ±35% is considered acceptable for initial estimates. However, this tolerance of inaccuracy should not be taken as a license to be sloppy or casual in early cost estimating.

In practice, each of the engineering disciplines has its unique approaches to cost estimating, and these approaches will often be captured by some useful *rules of thumb*, or heuristics or general guidelines, that are most relevant at the conceptual design stage. In civil engineering, for example, the *R. S. Means Cost Guide* provides cost estimates per square foot for the various elements in different kinds of construction projects. The *Richardson's Manual* offers similar information for chemical plant and petroleum refinery projects. On the other hand, for certain types of printed circuit board designs, costs per square inch may be relevant. In all of the disciplines, we would have to carefully consult with experienced professionals in order to successfully estimate costs; even at the more general levels we would need to make conceptual design choices.

Finally on cost estimation, we want to highlight the distinction between the cost of designing an artifact, on the one hand, and cost of manufacturing and distributing it, on the other. In many cases, the cost of design is a relatively insignificant share of the final project cost, such as in the case of a dam or other large structure. Notwithstanding that, however, most clients expect that the design team will have correctly estimated their own costs and budgeted them accurately. Thus, even when costing out the design activity, an effective design team will seek to accurately understand and control its costs.

8.2.4   Costing and Pricing

Finally, it should be noted that while costing is an important element in the *profitability* of a design, it is generally *not* a key factor in the *pricing* of the artifact. This seeming contradiction can be easily explained when we consider that gross profits (i.e., profits before taxes and other considerations) are simply the net of revenues minus costs. As such, costs are an important element in the profit equation. Revenues, on the other hand, are determined by the price charged for an item multiplied by the number of items sold. For most profit-maximizing firms, prices are not set on the basis of costs, but rather in terms of what the market is willing to pay. Some examples will illustrate this.

Consider a high-quality graphite tennis racket. This artifact may command a price in the hundreds of dollars when initially introduced. It is clear upon inspection, however, that the costs of such an object are nowhere near this figure. The materials for the racket may be as little as several dollars, the labor is virtually negligible, and the costs of technological developments are relatively modest when *amortized* (spread out) over the many thousands of sales of such a product. The distribution costs are clearly no higher for high-end rackets than for the $10 rackets found at discount stores. Since there is an obvious demand for these pricey rackets, it is clear that prices are set at the high end not because of the costs of production, but because there are customers willing to pay such high prices. Indeed, the role of marketing professionals on the design team usually includes identifying the attributes that will cause consumers to pay a premium for a design. This example also serves to highlight an aspect of reliability. A manufacturer may offer a virtually lifetime replacement guarantee if the costs of manufacture are so far below the selling price. Thus, the high-quality service of certain brands reflects a disparity between their pricing and their cost structure.

A similar example can be found in the airline industry. Here the provider of the service faces essentially the same costs regardless of whether a plane flies almost full or almost empty. This explains why airlines are willing to offer certain deeply discounted fares at some times, and almost no discounts at others (such as holidays). It also explains why the airlines are willing to invest heavily in modeling and tracking the wide variety of fare options they make available.

In some industries a convention has arisen to compensate designers or even providers of certain products on a "cost-plus" basis. For example, most large public works projects, such as highways or dams, are built on the basis of the costs of the contractor or designer plus an additional percentage as a profit allowance. While this is common practice in some cases, the norm in the private sector is to select prices to maximize profits, not to simply tack on a profit factor.

The point is that an engineering designer must control the costs of the design and ensure that the objectives of the client and users are realized. Beyond that, however, the ultimate profitability of an artifact may turn out to be beyond a designer's control, as in the cases of pet rocks and beanie babies.

8.3   DESIGN FOR RELIABILITY: HOW LONG WILL THIS THING WORK?

Most of us have a personal, visceral understanding of reliability and unreliability as a consequence of our own experience with everyday objects. Perhaps we say that the family car was very unreliable, or a good friend is a very reliable person—someone we can count on. While such informal assessments are acceptable in our personal lives, we need greater understanding and accuracy when we are functioning as engineering designers. Thus, in this section we will describe how engineers approach reliability, along with its sister concept, maintainability.

8.3.1   Reliability

To the engineer, reliability can be defined as "the probability that an item will perform its function under stated conditions of use and maintenance for a stated measure of the variate (time, distance, etc.)." This definition has a number of elements that warrant further comment. The first is that we can properly measure the reliability of a component or system only under the assumption that it has been or will be used under some particular set of usage and maintenance conditions. A second point to note is that the appropriate measure of use of the design, called the *variate*, may be something other than time. For example, the variate for a vehicle would be miles, while for a piece of vibrating machinery the variate would be the number of cycles. Third, we have to examine reliability in the context of the functions that we discussed in Chapter 5. This should serve to accentuate the care we should undertake in developing and defining the functions that a design must perform. Finally, we note that reliability is treated as a probability, and hence can be characterized by a distribution. In mathematical terms, this means that we can express our expectations of how reliable, or safe, or successful, we expect a product or a system to be in terms of a cumulative distribution function or a probability density function.

In practice, our use of a probabilistic definition enables us to consider reliability in the context of the opposite of success, that is, in terms of *failure*. In other words, we can frame our consideration of reliability in terms of the probability that a unit will fail to perform its functions under stated conditions within a specified window of time. This requires us to consider carefully what we mean by failure. British Standard 4778 defines a failure as "the termination of the ability of an item to perform a required function." This definition, while helpful at some level, does not capture some important subtleties that we, as designers, must keep in mind. The definition doesn't capture the many kinds of failures that can afflict a complex device or sytem, their degree of severity, their timing, or their effect on the performance of the overall system.

For example, we find it useful to distinguish between when a system fails and how it fails. If the item fails when in use, the failure can be characterized as an *in-service failure*. If the item fails, but the consequences are not detectable until some other activity takes place, we refer to that as an *incidental failure*. If the nature of a failure of some function is such that the entire system of which the item is part also fails, we call that a *catastrophic failure*. If, for example, our car breaks down while we're on a trip and needs a repair in order for us to complete the trip, we would call that an in-service failure. An incidental failure might be some part that our favorite mechanic suggests we replace during the routine servicing of our car. A catastrophic, accident-causing failure might follow from the failure of a critical part of the car while we are driving at freeway speeds. Each type of failure has its own consequences for the users of the designed artifact, and so must be considered carefully by designers.

We often specify reliability in part by using measures such as the mean time between failures (MTBF), or miles per in-service failure, or some other variate or metric. However, we should note that framing the definition of reliability in terms of probabilities can give us some insight into the limitations inherent in such measures. Consider the two failure distributions shown in Figure 8.1. These two reliability probability distributions have the same mean (or average), that is, $\text{MTBF}_a = \text{MTBF}_b$, but they have very different degrees of dispersion (typically measured as the variance or standard deviation) about that mean. If we are not concerned with both mean and variance, we may wind up choosing a design alternative that is seemingly better in terms of MTBF, but much worse in terms of variance. That is, we may even choose a design for which the MTBF is acceptable, but for which the number of early failures is unacceptably high.

One of the most important reliability considerations for a designer is knowing how the various parts of the design come together and what the impact is likely to be if any one part does fail. Consider, for example, the conceptual sketch of a design, shown in Figure 8.2, that we refer to as a *series system*. It is a chain of parts or elements. The failure of any one of the elements would break the chain, which in turn will cause the system to fail. Just as a chain can be no stronger than its weakest link, a series system can be no more reliable than its most unreliable part. In fact, the reliability—or probability that the system will function as designed—of a series system is given by:

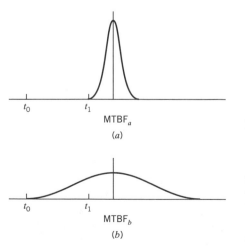

t_0 t_1
MTBF$_a$
(a)

t_0 t_1
MTBF$_b$
(b)

Figure 8.1 Failure distributions (also called probability density functions) for two different components. Note that both curves have the same value of MTBF, but that the dispersions of possible failures differ markedly. The second design (sketch b) would be viewed as less reliable because more failures would occur during the early life of the component (i.e., during the time interval $t_0 \leq t \leq t_1$).

Figure 8.2 This is a simple example of a *series system*. Each of the elements in the system has a given reliability. The reliability of the system as a whole can be no higher than that of any one of the parts because the failure of any one part will cause the system to cease operating. What is the reliability of this system as calculated with Eqn. (8.2)?

$$R_s(t) = R_1(t) \bullet R_2(t) \bullet \cdots \bullet R_n(t)$$

or

$$R_s(t) = \prod_{i=1}^{n} R_i(t) \tag{8.2}$$

where $R_s(t)$ is the reliability of the entire series system, $R_i(t)$ is the reliability (or probability of successful performance) of each of the individual components in the system, and $\prod_{i=1}^{n}$ is the product function. We see from Eqn. (8.2) that the overall reliability of a series system is equal to the product of all of the individual reliabilities of the elements or parts within the system. This means that if any one component has low reliability, such as the proverbial weak link, then the entire system will have low reliability, and the chain will break.

Designers have long understood the value of redundancy as an approach to dealing with the weakest-link phenomenon. A *redundant system* is one in which some or all of the parts have backups or replacement parts that can subsitute for them in the event of failure. Consider the conceptual sketch shown in Figure 8.3 wherein three parts or elements are arranged in what we call a *parallel system*. In this simple case, each of the components must fail in order for the system to fail. The reliability $R_p(t)$ of this entire parallel system is given by:

$$R_p(t) = 1 - \left[\left(1 - R_1(t)\right) \bullet \left(1 - R_2(t)\right) \bullet \cdots \bullet \left(1 - R_n(t)\right) \right]$$

or

$$R_p(t) = 1 - \prod_{i=1}^{n} \left[1 - R_i(t) \right] \tag{8.3}$$

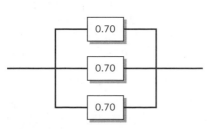

Figure 8.3 This is a simple example of a *parallel system*. Note that every one of the components must fail in order for the system to cease working. While such a system has high reliability, it is also quite expensive. Most designers seek to incorporate such redundancy when necessary, but look for other solutions wherever possible. How does the reliability of this parallel system, as calculated with Eqn. (8.3), compare with that of the series system in Figure 8.2?

We see from Eqn. (8.3) that the reliability of this parallel system (i.e., the probability that the parallel system will operate successfully) is now such that if any one of the elements functions, the system will still function. Parallel systems have obvious advantages in terms of reliability, since all of the redundant or duplicate parts must fail in order for the system to fail. Parallel systems are also more expensive, since many duplicate parts or elements are included only for contingent use, that is, they are used only when another part has failed. For this reason, we must must carefully weigh the consequences of failure of a part against failure of the system, along with costs attendant to reducing the likelihood of the failure. In most cases designers will opt for some level of redundancy, while allowing other components to stand alone. For example, a car usually has two headlights, in part so that if one fails the car can continue to operate safely at night. The same car will usually only have one radio, since its failure is unlikely to be catastrophic. The mathematics of combining series and parallel systems is beyond our scope, but we clearly have to learn and use them when we are designing systems that have any impact on the safety of users.

Designers can consider modes of failure and develop estimates of reliability only if they truly understand how components might fail. This understanding can be gained in a number of ways, including by performing experiments, analyzing the statistics of prior failures, or by carefully modeling the underlying physical phenomena. Absent deep experience in understanding component failure, designers should consult experienced engineers and designers, users, and the client in order to ascertain that the level of reliability being designed for is appropriate. In many cases, the experiences of others will allow a designer to answer reliability questions without performing a full set of experiments. For example, during the design of a physical system, the appropriateness of different kinds of materials for various purposes can be discussed with experienced materials engineers, while properties such as tensile strength and fatigue life are well documented in the engineering literature.

8.3.2 Maintainability

Our understanding of reliability also leads us to conclude that many of the systems that we design do fail if they are used without being maintained, and that they may need some amount of repair even if they are properly maintained. This fact of life has lead engineers to consider how best they might design things so that the necessary maintenance can be performed effectively and efficiently. Maintainability can be defined as "the probability that a failed component or system will be restored or repaired to a specific condition within a period of time when maintenance is performed within prescribed procedures." As we did with our definition of reliability, we can learn some things from this definition. The first is that maintainability depends upon a prior specification of the condition of the part or device, and on any maintenance or repair actions, which are part of the designer's responsibilities. A second point is that maintainability is concerned with the time needed to return a failed unit to service. As such, many of the same concerns regarding inappropriate use of a single measure, for example, the mean time to repair (MTTR), come into play.

Designing for maintainability requires that the designer take an active role in setting goals for maintenance (such as times to repair), and in determining the specifications for maintenance and repair activities in order to realize these goals. This can take a number of forms, including selecting parts that are easily accessed and repaired, providing redundancy so that systems can be operated while maintenance continues, specifying preventive or predictive maintenance procedures, and indicating the number and type of spare parts that should be held in inventories in order to reduce downtime when systems fail. There are costs and consequences for the designed system in each of these design choices. If, for example, a system is designed to have increased redundancy in order to limit downtime while maintenance is ongoing, as is the case in our air traffic control system, the attendant capital costs can be very large. Similarly, the cost of carrying inventories of spare parts can be quite high, especially if failures are rare. One strategy that has been increasingly adopted in many industries is to work toward making parts standard and components modular. Then spare parts inventories can be used more flexibly and efficiently, and components or subassemblies can be easily accessed and replaced. Any removed subassemblies can be repaired while the repaired system has been returned to service.

In view of what we've just said, it seems clear that design teams must ensure that they take active steps in the design process to meet the objective of high maintainability, if that has been established as a significant design goal. As such, a design team should ask itself what maintenance actions (e.g., service) will reduce failures, what elements of the design will support early detection of problems or failures (e.g., inspection), and what elements will ease the return of failed items to use (e.g., repair). While no one would intentionally design systems in order to make maintenance more difficult, the world is fraught with examples in which it is difficult to believe otherwise, including a new car in which the owner would have to remove the dashboard just to change a fuse!

8.4   NOTES

Section 8.1: This section draws heavily upon (Pahl and Beitz 1996) and (Ulrich and Eppinger 1995). In particular, the six-step process is a direct extension of a five-step approach in (Ulrich and Eppinger 1995), with the iteration made explicit. Our discussion of DFA is adapted from (Dixon and Poli 1995) and (Ullman 1997); rules of assembly are cited in many places but are generally derived from (Boothroyd and Dewhurst 1989). The analogy of the BOM to a recipe is taken from (Schroeder 1993).

Section 8.2: Our review of basic concepts derives from (Riggs and West 1986), but there are any number of excellent texts in engineering economics that can be used to go further with these topics. The cost-estimating material draws upon classnotes prepared by our colleague Donald Remer for his cost-estimating course and on (Oberlander 1993). A view of engineering costing that is based more on accounting may be found in (Riggs 1994). The relationship between pricing and costs is discussed in (Nagle 1987) and (Phlips 1985).

Section 8.3: The definition of reliability comes from U.S. Military Standards Handbook 217B (MIL-STD-217B 1970) as quoted in (Carter 1986). The failure discussion draws heavily from (Little 1991). There are a number of formal treatments of reliability and the associated mathematics, including (Ebeling 1997) and (Lewis 1987). The definition of maintainability is from (Ebeling 1997). The distinction between service, inspection, and repair is from (Pahl and Beitz 1996).

8.5 EXERCISES

8.1 If you were asked to design a product for recyclability, how would you determine what that meant? In addition, what sorts of questions should you be prepared to ask and answer?

8.2 How might DFA considerations differ for products made in large volume (e.g., the portable electric guitar) and those made in very small quantities (e.g., a greenhouse)?

8.3 Your design team has produced two alternative designs for city buses. Alternative A has an initial cost of $100,000, estimated annual operating costs of $10,000, and will require a $50,000 overhaul after five years. Alternative B has an initial cost of $150,000, estimated annual operating costs of $5,000, and will not require an overhaul after five years. Both alternatives will last ten years. If all other vehicle performance characteristics are the same, determine which bus is preferable using a discount rate of 10%.

8.4 Would the decision reached in Exercise 8.3 change if the discount rate was 20%? What would happen if, instead, the discount rate was 15%? How do the resulting cost figures influence your assessments of the given cost estimates?

8.5 Your design team has produced two alternative designs for greenhouses in a developing country. Alternative A has an initial cost of $200 and will last two years. Alternative B has an initial cost of $1,000 and will last ten years. All other things being equal, determine which greenhouse is more economical with a discount rate of 10%. Which other other factors can influence this decision?

8.6 What is the reliability of the system portrayed in Figure 8.2?

8.7 What is the reliability of the system portrayed in Figure 8.3? How does this result compare with that of Exercise 8.6? Why?

8.8 On what basis would you choose between a single system, all of whose parts are redundant, and two copies of a system that has no redundancy?

Chapter 9

Ethics in Design

Isn't design really just a technical matter?

If we think back to our early discussions of design in Chapter 1, as brief and as general as they were, we see that design is very much a human endeavor; it's not done in a vacuum. In addition to being an intellectual activity, *design is a social activity*. Design involves people in many different ways, ranging from the interactions among members of a design team, through the relationships between designers, clients, and manufacturers, to the ways that purchasers of designed devices use them in their lives. Because design touches so many facets of people's daily lives, we have to consider how people interact with each other as they create and then use a design. Further, doing a design means accepting responsibility for creating a design for people to use. Therefore, we have to consider ethics and ethical behavior as part of our examination of how a design is created and then used.

9.1 ETHICS: RECONCILING CONFLICTING OBLIGATIONS

Inasmuch as words like ethics, morals, obligations, and duty, among others, carry baggage of their own, we will start this discussion as we have others, with some dictionary definitions. First, the word *ethics*:

> **ethics 1** the discipline dealing with what is good and bad and with moral duty and obligation **2 a:** a set of moral principles or values **b:** a theory or system of moral values **c:** the principles of conduct governing an individual or group

And since it is referenced so often in the definition of ethics, the word *moral*:

> **moral 1 a:** of or relating to principles of right or wrong in behavior
> **b:** expressing or teaching a conception of right behavior

One straightforward interpretation of these two definitions is that beyond defining a discipline or field of study, ethics refers to a set of guiding principles or a system that people can use to help them behave rightly or well. But, didn't we learn right and wrong from our parents? Perhaps we also learned it as a set of religious tenets—that could have happened in religious traditions that emphasize faith in God (e.g., Christianity, Judaism, and Islam) or those that stress faith in a *right path* (e.g., Buddhism, Confucianism, and Taoism). Either way, we know about honesty and integrity, and about the injunction to treat others as we would want to be treated ourselves.

If we do know these things, why do we need another (external) set of rules or agencies to teach us what we already know? If we don't, and the law doesn't keep us in line, what's the use of a set of ethical principles? The truth is that the kinds of lessons we learn at home, in school, and in churches, mosques, synagogues, and temples, may not provide enough explicit guidance about many of the situations we face in life, especially in our professional lives. In addition, given the diversity and complexity of our society, we are likely better off if we have some standards of professional behavior that are universally agreed upon, across all of our traditions and individual upbringings. (If everyone's individually-learned lessons were sufficient, our dependence on laws and lawyers would be drastically—and happily—diminished!)

One reason that our professional lives are more complicated is that part of our newfound responsibilities involve obligations to many stakeholders, some of whom are obvious (e.g., clients, users, perhaps the general public) and some of whom are not (e.g., some government agencies, professional societies). We will elaborate these obligations further in Sections 9.2–9.4, but for now we note that these obligations often conflict. For example, a client may want one thing, while a group of people affected by a design result may want something altogether different. Still further, those affected by a design may not even know how they are being affected until *after* the design is complete and some final action has been taken or the design has been implemented. We will see a notable version of this clash of interests and obligations in our case study in Chapter 10. In the meantime, consider the following scenario.

Imagine you are a mining engineer in the United States, late in the 19th century. You have been engaged by the owner of a mine to design a new shaft extension, and as part of that design task you survey the mine. As you execute that commission you find out that part of the mine runs under someone else's property. Aren't you obligated (i.e., isn't it your duty) to simply complete the survey and the design for the mine owner who engaged you and who is paying you, and then go on to your next professional engagement?

Perhaps you feel a sneaking suspicion that the mine owner has not notified the landowner that his mineral rights are being excavated out from underneath him. Is there something you should do about that? If so, what? What compels you to do something? Is it personal morality? Is there a law? To whom are you responsible? And in what ways are you responsible?

The chain of questions just started can easily be lengthened, and the situation can easily be made more complicated. For example, suppose the mine was the only mine in town, and thus its owner basically controlled your livelihood, as well perhaps those of most of the town residents? For another example, suppose you were doing this late in the 20th century and you find out that the mine runs under, and perilously close to, a school. Does that change things?

This little story highlights some of the many different actors and obligations that could arise in an engineering design project. It also happens to be something like a case study because scenarios such as this actually did happen in the latter part of the 19th century and on into the 20th. It was just such situations that gave great impetus both to the formation of professional societies and to the development and promulgation of codes of ethics by these same societies. The societies were formed and the ethical codes proclaimed in part as a form of protection for the individual members.

Over time, the professional societies also came to promulgate standards for design endeavors, such as those we described in Section 7.3, and to undertake many other kinds of activities, including providing forums for reporting new research and innovations in practice. However, it is not our intent to provide a history of professional societies. Rather, we simply want to make the point that professional engineering societies play an important role in setting not only design standards, but also ethical standards for designers and engineers. These ethical standards clearly speak to the various and often conflicting obligations that an engineer must meet. The societies also provide mechanisms for helping engineers to deal with and resolve conflicting obligations, and they provide means for, when asked to, investigating and evaluating ethical behavior.

To return to our case study, we can see that were you the designer, you might have to try to meet a range of obligations. Certainly you owe the mine owner a conscientious and professional shaft design, including within that a reliable and unassailable survey of the mine itself. However, when you discover the fact of the mine being under someone else's land, you might feel it is your duty to query the mine owner about what he's doing, to approach the landowner and see what she knows, or to go to some external authority either to ask questions or raise allegations. In the later version of our example, where a school may be underneath the mine, more people (i.e., schoolchildren and their parents) and agencies (i.e., the school board) would be involved, as a result of which you might feel still more obligations. In any event, in either scenario, as the designer you would likely feel pulled or obligated toward several people or groups. That is why there is a need for some structure or guidelines to assist you in framing a response and formulating a plan to deal with your concerns.

The situation described in our example is far from unique. One of the better-known instances of engineers who faced difficult and noteworthy ethical dilemmas is the group of engineers that tried, unsuccessfully, to delay the launch (on January 28, 1986) of the space shuttle *Challenger*. While severe doubt was being expressed by some engineers about the safety of the *Challenger*'s O-rings because of the cold weather before the flight, the upper management of Morton-Thiokol, the company that made the *Challenger*'s booster rockets, and NASA approved the launch. These managers determined that their respective concerns about Morton-Thiokol's image and the stature and visibility of NASA's shuttle program outweighed the judgments of the engineers closest to the design of the booster itself. The Morton-Thiokol engineers who publicized the dismissal of their recommendation not to launch were engaged in *whistleblowing*, wherein someone "blows the whistle" in order to stop or prevent a faulty decision made within a company, agency, or some other institution.

Such instances of whistleblowing are, unfortunately, neither new nor entirely unique. One of the earlier cases that became nationally famous was that of an industrial engineer, Ernest Fitzgerald, who blew the whistle on major cost overruns in the procurement of the Air Force's giant C–5A cargo plane. The Air Force was sufficiently displeased with Fitzgerald's actions that it took classic bureaucratic actions to keep him from further work on the C–5A: It "lost" his Civil Service tenure and then reconstructed that part of the bureaucracy in which Fitzgerald worked so as to eliminate his position! After a long, expensive, and arduous legal battle, Fitzgerald earned a substantial settlement for wrongful termination and was reinstated to his position.

While these stories are to some extent discouraging, they can also be seen as showing heroic human behavior under very trying circumstances. Even more to our point, however, is the fact that our examples show quite clearly how "doing what's right" can be perceived quite differently within an organization. An engineer or designer may well be faced with just the kind of clash or conflict of obligations that we think lies at the crux of any discussion of engineering ethics. If that happens, to whom can the designer or engineer turn for help?

Part of that answer lies, in fact, in the roots of the engineer's own understanding of her ethics, that is, in herself, her family, her own personal beliefs. Another part of the answer lies in the support of immediate professional colleagues and peers, as these have been seen to be very effective in many instances in both righting the perceived wrong and in sustaining the whistleblower. Then, of course, there are the professional societies.

Most of the professional engineering societies have established and published codes of ethics. As a representative sample, we show the codes of ethics of the American Society of Civil Engineers (ASCE) in Figure 9.1 and of the Institute of Electronics and Electrical Engineers (IEEE) in Figure 9.2. While both of these codes highlight integrity and honesty, they also appear to value certain kinds of behavior differently. For example, the ASCE code enjoins its members from competing unfairly with others, a subject not mentioned at all by the IEEE. Similarly, the IEEE specifically calls for its members to "fairly treat all persons regardless of such factors as race, religion, gender . . ." There are other differences as well, for example, in the styles of language. The ASCE seems to present a stern set of injunctions about what engineers "shall" do, while the IEEE phrases its code as a set of commitments to undertake certain behaviors.

Notwithstanding these differences, what both codes of ethics do is set out guidelines or standards of how to behave with respect to: clients (e.g., ASCE's

Figure 9.1 The code of ethics of the American Society of Civil Engineers (ASCE), dated January 1, 1977. It is similar, although not identical, to the code adopted by the IEEE that is displayed in Figure 9.2.

ASCE CODE OF ETHICS

Fundamental Principles

Engineers uphold and advance the integrity, honor and dignity of the engineering profession by:

1. using their knowledge and skill for the enhancement of human welfare and the environment;
2. being honest and impartial and serving with fidelity the public, their employers and clients;
3. striving to increase the competence and prestige of the engineering profession; and
4. supporting the professional and technical societies of their disciplines.

Fundamental Canons

1. Engineers shall hold paramount the safety, health and welfare of the public and shall strive to comply with the principles of sustainable development in the performance of their professional duties.
2. Engineers shall perform services only in areas of their competence.
3. Engineers shall issue public statements only in an objective and truthful manner.
4. Engineers shall act in professional matters for each employer or client as faithful agents or trustees, and shall avoid conflicts of interest.
5. Engineers shall build their professional reputation on the merit of their services and shall not compete unfairly with others.
6. Engineers shall act in such a manner as to uphold and enhance the honor, integrity, and dignity of the engineering profession.
7. Engineers shall continue their professional development throughout their careers, and shall provide opportunities for the professional development of those engineers under their supervision.

"as faithful agents or trustees"); the profession (e.g., IEEE's "assist colleagues and co-workers in their professional development"); the public (e.g., ASCE's "shall issue public statements only in an objective and truthful manner"); and the law (e.g., IEEE's "reject bribery in all its forms"). These kinds of standards are designed to lay out rules of the road for dealing with conflicting obligations, including for the task of assessing whether these conflicts are "only" of perception or of a substantial, "real," and potentially damaging nature.

We close with two further notes about the professional societies and their codes. First, it should be noted that the differences in the codes reflect less—

Figure 9.2 The code of ethics of the Institute of Electronics and Electrical Engineers (IEEE), dated August 1990. How does the IEEE code of ethics differ from that adopted by the ASCE that is displayed in Figure 9.1?

IEEE CODE OF ETHICS

We, the members of the IEEE, in recognition of the importance of our technologies in affecting the quality of life throughout the world, and in accepting a personal obligation to our profession, its members and the communities we serve, do hereby commit ourselves to the highest ethical and professional conduct and agree:

1 to accept responsibility in making engineering decisions consistent with the safety, health, and welfare of the public, and to disclose promptly factors that might endanger the public or the environment;

2 to avoid real or perceived conflicts of interest whenever possible, and to disclose them to affected parties when they do exist;

3 to be honest and realistic in stating claims or estimates based on available data;

4 to reject bribery in all its forms;

5 to improve the understanding of technology, its appropriate application, and potential consequences;

6 to maintain and improve our technical competence and to undertake technological tasks for others only if qualified by training or experience, or after full disclosure of pertinent limitations;

7 to seek, accept, and offer honest criticism of technical work, to acknowledge and correct errors, and to credit properly the contributions of others;

8 to treat fairly all persons regardless of such factors as race, religion, gender, disability, age, or national origin;

9 to avoid injuring others, their property, reputation, or employment by false or malicious action;

10 to assist colleagues and co-workers in their professional development and to support them in following this code of ethics.

---

indeed, much less—differences in their views of the importance of ethics than they do the different styles of engineering practice in the various disciplines. For examples, many civil engineers not employed in a government agency work in small, "Mom and Pop" consulting companies that are people-intensive rather than capital-intensive. These firms obtain much of their work through public, competitive bidding. Electrical engineers, on the other hand, more often than not work for large corporations that sell products more than they do services, one result of which is that they tend to be capital-intensive and have significant manufacturing operations. Such differences in their practices produce different cultures, and hence, different statements of ethical standards.

The second point is that, notwithstanding their promulgation of codes of ethics, the professional societies have not always been seen as active and visible protectors of whistleblowers and other professionals who raise concerns about specific engineering or design instances. While this situation is improving—steadily if slowly—many engineers still find it difficult to look to their societies, and especially their local branch sections, for first-line assistance and support in times of need. Of course, as we all make ethical behavior a higher priority, the need for such support will lessen, and its ready availability will surely increase.

9.2   OBLIGATIONS MAY START WITH THE CLIENT . . .

We start by considering our various obligations to the client or, perhaps, to an employer. There's little doubt that we, as designers or engineers, owe our client or our employer a professional effort at solving a design problem, by which we mean being technically competent, conscientious, thorough, and that we should undertake technical tasks only if we are properly "qualified by training or experience." We also need to avoid any conflicts of interest, and to disclose any that might exist. And we must serve our employer by being "honest and impartial" and by "serving with fidelity . . . " Most of these obligations are clearly delineated in codes of ethics (e.g., compare the quotes with Figures 9.1 and 9.2), but there's at least one curious obligation on this list. What precisely does it mean to serve with "fidelity"?

A thesaurus would tell us that fidelity has several synonyms, including constancy, fealty, allegiance, and loyalty. Thus, one implication we can draw from the ASCE code of ethics (at least) is that we should be loyal to our employer or our client. This suggests that one of our obligations is to look out for the best interests of our client or employer and to maintain a clear picture of those interests as we do our design work. But loyalty is a very touchy matter; it is not a simple, one-dimensional attribute. In fact, clients and companies earn the loyalty of their consultants and staffs in at least two ways. One, which is called *agency-loyalty*, simply derives from the nature of any contracts between the designer and his client (e.g., "work for hire") or his employer (e.g., a "hired worker"). As it is dicated by contract, agency-loyalty is clearly obligatory for the designer. The second kind of loyalty, which is called *identification-loyalty*, is perhaps more likely to be viewed as optional. It stems from the designer identifying with the client or company because he admires its goals or sees its behavior as mirroring his own values. To the extent that identification-loyalty is an optional obligation, it can be earned by clients and companies only if they demonstrate reciprocal loyalty to their designers.

Agency-loyalty provides one reason for us to maintain a "design notebook" to document our design work. As we have noted earlier, keeping such a record is good design practice because it is very useful for the real-time tracking and recapitulation of our thinking as we move through different stages of the design process. However, a dated design notebook also provides the legal basis for documenting how new, patentable ideas were developed. Such documentation would be essential to our employer or client if a patent application is in any way challenged. Further, the intellectual work done in the creation of a design is itself the intellectual property of the client or the employer, and this is typically specified in contracts and employment agreements. There is no reason that a client or an employer cannot share the rights to that intellectual property with its creators, but the fundamental decisions about the ownership of the property generally belong to the client or owner. It is important for a designer to keep that in mind, and also to document any separate, private work that she is doing, just to avoid any confusion about who owns any particular piece of design work.

The optional nature of identification-loyalty provides the fertile ground for a clash of obligations, because it is here that other loyalties have the space to make themselves felt. As we will discuss further in Section 9.3.1, modern codes of ethics normally have language articulating some form of obligation to the general health and welfare of the public. For example, in the ASCE code of ethics (viz., Figure 9.1) it is suggested that engineers work toward enhancing both human welfare and the environment, and that they "shall hold paramount the safety, health, and welfare of the public . . ." Similarly, the IEEE code (viz., Figure 9.2) suggests that its members commit to "making engineering decisions consistent with the safety, health, and welfare of the public . . ." These are clear calls to engineers to identify other loyalties to which they should feel allegiance. There's little doubt that it was just such divided loyalties that emerged in the many reported cases of whistleblowing.

In the case of the explosion of the *Challenger*, those who argued against its launching felt that lives would be endangered. They clearly placed a higher value on the lives being risked than they did on the loyalties demanded by Morton-Thiokol (i.e., to secure its place as a government contractor) and by NASA (i.e., to its ability to successfully argue for the shuttle program before Congress and before the public). Similar conflicting loyalties have emerged for engineers, and others, as toxic waste sites are cleaned up under the Super Fund program of the Environmental Protection Agency (EPA). In many instances, employees have felt that they needed to look out for their own companies, sometimes because they felt the companies should not be penalized for doing what was once legal, in others because they were pressured to do so by peers and bosses and by the possible loss of their jobs. There are also cases in which engineers were apparently willing, or at least able, to rank their loyalties to their companies first, to the point where falsified emission test data were reported to the government (by engineers and managers at the Ford Motor Company) or parts known to be faulty were delivered to the Air Force (by engineers and managers at the B. F. Goodrich Company).

Some of the time, an apparent disloyalty to a company or an organization may be, in a longer term, an act of greater and succesfully merged loyalties. For example, when the Ford Pinto was initially being designed, some of its engineers wanted to perform some crash tests that were not then called for in the relevant U.S. Department of Transportation regulations. The managers charged with developing the Pinto felt that such tests could not benefit the program and, in fact, might only prove to be a burden. Why, after all, should we run a test that's not required, only to risk failing that test? The designers who proposed the tests were viewed as disloyal to Ford Motor Company and the Pinto program. In fact, what happened was that the placement of the drive train and the gas tank resulted in fiery crashes, lives lost, and some major public relations and financial headaches for Ford. Clearly, Ford would have been better off in the long run to have conducted the tests, so that the engineers who proposed them could be said to have been looking out for the company's interests in the long term.

If there's a single point that emerges from the discussion thus far, it is that ethical issues are not resolved by identifying a *single* obligation. Indeed, were choices and issues so clearly categorized, ethics wouldn't be a problem because there wouldn't be any choices to make.

9.3   ... BUT WHAT ABOUT THE PUBLIC AND THE PROFESSION?

Now we want to tell a story that shows, among other things, that things can turn out well in bad situations when people behave well. In fact, to start with the ending, the protagonist-hero of this story has been quoted as saying, "In return for getting a [professional engineering] license and being regarded with respect, you're supposed to be self-sacrificing and look beyond the interests of yourself and your client to society as a whole. And the most wonderful part of my story is that when I did it nothing bad happpened."

Our hero is William J. LeMessurier (which he pronounces as "LeMeasure") of Cambridge, Massachusetts, one of the most highly regarded structural engineers and designers in the world. He served as the structural consultant to a noted architect, Hugh Stubbins, Jr., for the design of a new headquarters building in New York for Citicorp, the parent corporation of one of the nation's largest banks, CitiBank. Completed in 1978, the 59-story Citicorp Center is one of the most dramatic and interesting skyscrapers in a city filled with some of the world's great buildings (see Figure 9.3). In many ways, LeMessurier's conceptual design for Citicorp resembles that of other striking skyscrapers in that it used a *tube* concept in which the building is designed as a tall, hollow tube that has a comparatively rigid or stiff tube wall. (The structural engineering terminology for this is that the tube's main lateral stability elements are located at the outer perimeter and tied together at the corners.) One similar design is Fazlur Kahn's John Hancock Center in Chicago (see Figure 9.4). Here the outer "tube" or "main lateral stability elements" can be seen in the multistory diagonal elements that are joined to large columns at the corners. Kahn's design benefited from an architectural decision to deliberately expose the details of the tube, perhaps to illustrate the famous dictum that form follows funtion (which was actually pronounced by Louis Sullivan, the noted Chicago architect who was also Frank Lloyd Wright's mentor).

LeMessurier's design of Citicorp Center was innovative in several ways. One, not directly visible from the outside, was the inclusion of a large mass, floating on a sheet of oil, within the triangular roof structure. This element was added as a damper in order to reduce or damp out any oscillations the building might undergo due to wind forces. Another innovation was LeMessurier's adaptation

Figure 9.3 One view of the 59-story Citicorp Center, designed by architect Hugh Stubbins, Jr., with William J. LeMessurier serving as structural consultant. One of this building's notable features is that it rests on four massive columns that are placed at the midpoints of the building's sides, rather than at the corners. This enabled the architects to include under the Citicorp's sheltering canopy a new building for St. Peter's Church. (Photo by Clive L. Dym.)

of the tube concept to an unusual situation. The land on which the Citicorp Center was built had belonged to St. Peter's Church, and the church itself occupied an old (dating from 1905) and decaying Gothic building on the lot's west side. As part of the sale of the building lot to Citicorp, St. Peter's gained a new church for itself, to be erected "under" the Citicorp skyscraper. In order to manage this, LeMessurier moved the "corners" of the building to the midpoints of each side (see Figures 9.3 and 9.5). This enabled the creation of a large space for the new church because the office tower itself could be cantilevered out over the church, at a height of some nine stories. Looking at the sidewalls of the tube— and here we have to peel off the building's skin because architect Stubbins did not want the structures exposed, as they were in the Hancock tower—we see that the wall rigidity comes from large triangular elements made up of diagonal and horizontal elements. The triangles are set up so they all connect at the midpoints of the sides. Thus, LeMessurier's triangles serve here the same basic purpose as Kahn's large X-frames.

The ethics problem arose not long after the building was completed and occupied. LeMessurier received a call from an engineering student in New Jersey who was told by a professor that the building's columns had been put in the wrong place. In fact, LeMessurier was very proud of his idea of placing the columns at the midpoints. He explained to the student how the 48 diagonal

Figure 9.4 The 102-story John Hancock Center, designed by the architectural firm of Skidmore, Owings and Merrill, with Fazlur Kahn serving as structural engineer. Note how the exposed diagonal and column elements make up the tube that is the building's underlying conceptual design. (Photo by Clive L. Dym.)

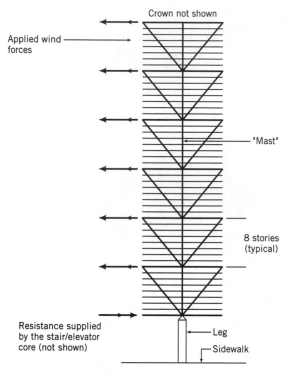

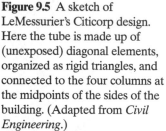

Figure 9.5 A sketch of LeMessurier's Citicorp design. Here the tube is made up of (unexposed) diagonal elements, organized as rigid triangles, and connected to the four columns at the midpoints of the sides of the building. (Adapted from *Civil Engineering*.)

braces he had superposed on the midspan columns added great stiffness to the building's tube framework, particularly with respect to wind forces. The student's call and questions also intrigued LeMessurier enough to go back to his original design and calculations to see just how strong the wind bracing system would be in a case that was not examined under then-current practice and building codes. Practice at the time called for wind force effects to be calculated when the wind flow hit a side of a building dead on a face, that is, normal to the building sides. What had not been called for before this case was a calculation of the effect on a building of a *quartering wind*, under which the wind hit the building on a 45-degree diagonal, so that wind pressure was then distributed over the two immediately adjacent faces (see Figure 9.6). It turned out that in this case, some of the diagonals are unstressed, while others are doubly loaded, which led to calculated strain increases of 40 percent. Normally, even this increase in strain (and stress) would not have been a problem because of the basic assumptions under which the entire system was designed.

However, to LeMessurier's further discomfort, he found out just a few weeks later that the actual connections in the finished diagonal bracing system were not the high-strength welds he had stipulated. Rather, it turned out that they were bolted because the steel fabricator, Bethlehem Steel, had determined and suggested to LeMessurier's New York office that bolts would be more than strong enough and, at the time, significantly cheaper. This choice of bolts was sound and quite correct professionally. However, to LeMessurier the bolts meant that

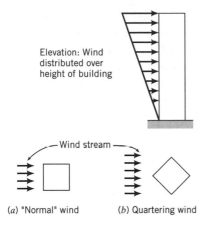

Elevation: Wind
distributed over
height of building

—Wind stream—

(a) "Normal" wind   (b) Quartering wind

Figure 9.6 A sketch of how wind forces are seen on buildings. (a) This is the standard case of wind streams being normal or perpendicular to the building faces. It was the only one called for in design codes at the time the Citicorp center was designed. (b) This is the case of quartering winds, in which case the wind stream comes along a 45-degree diagonal and thus simultaneously applies pressure along two faces at once.

the margin of safety against forces due to a quartering wind—which, again, structural engineers were not then called upon to consider—was not as large as he would have liked. (It is interesting to note that while New York City's building code did not then require that quartering winds be considered in building design, Boston's code did—and had since the 1950s!)

In fact, spurred by his new calculations and the news about the bolts, as well as by learning of some other detailed design assumptions made by engineers in his New York office, LeMessurier retreated to the privacy of his summer home on an island in a lake in Maine to carefully review all of the calculations and changes, and their implications. After doing a member-by-member calculation of the forces and reviewing the weather statistics for New York City, LeMessurier determined that the Citicorp Center had a probability of being subjected to winds that would produce a catastrophic failure once every sixteen years. Thus, in terminology used by meteorologists to describe both winds and floods, the Citicorp Center would fail in a sixteen-year storm—but it had supposedly been designed to withstand a fifty-year storm. What was LeMessurier to do?

In fact, he considered a range of options, including, reportedly, that of driving into a freeway bridge abutment at high speed. He also considered remaining mute and silent, and he tried to reassure himself that his innovative rooftop mass damper actually did reduce the probabilities of such failure to the fifty-year level. On the other hand, if the power went out, the mass damper wouldn't be there to help. What did LeMessurier actually do?

He started by trying to contact the architect, Hugh Stubbins, who was away on a trip. He then reached Stubbins' lawyer, after which he talked first with his own insurance carrier and then with the principal officers of Citicorp, one of whom studied engineering before choosing to become a banker. While some consideration was initially given to evacuating the building, especially since hurricane season was just over the horizon, it was decided to redesign all of the connections that were at risk and to retroactively fix them. Steel plate Band-Aids, two inches thick, would be welded onto each of 200 bolted connections. However, there were some very interesting implementation problems, only some of which do we mention here. The building's occupants had to be informed without being alarmed because the repair work would go on every night for two months or more. The public had to be informed why the bank's new flagship headquarters suddenly needed immediate modifications. (In fact, the entire process seemed to be open and attentive to public concerns.) Skilled structural welders, who were in short supply, had to be found, as did an adequate supply of the right grade of steel plate. Discreet, even secret, evacuation plans had to be put in place, just in case an unexpectedly high wind did show up while the repairs were being made. Also central to resolving the problem, New York City's Building Commissioner and the Department of Buildings and its inspectors had to be brought into the loop. They had to be informed of the problem and of its proposed solution, and they had to agree to the inspection of that solution. All in all, a dazzling array of problems, institutions, and, of course, personalities.

In the end, the steel Band-Aids were applied and the entire business was completed professionally, with no finger-pointing or public assignments of blame. LeMessurier, who had thought his career might precipitously end, came away with perhaps even greater stature, occasioned by his willingness to candidly face up to the problem and to propose a realistic, carefully crafted solution. In the words of one of the engineers involved in implementing LeMessurier's solution, "It wasn't a case of 'We caught you, you skunk.' It started with a guy who stood up and said, 'I got a problem, I made the problem, let's fix the problem.' If you're gonna kill a guy like LeMessurier, why should anybody ever talk?"

As we have said before, this is a case where everyone involved behaved well. In fact, it is to everyone's credit that all of the actors behaved very well, to a very high standard of professionalism and understanding. It is, therefore, a case that we can study with pleasure, particularly as engineers. It also is a case that could have gone other ways, so we will close our discussion by posing a few questions that you, the reader, might face were you in LeMessurier's position:

- Would you have "blown the whistle," or not?
- What would you have done if you determined that the revised probability of failure was higher (i.e., worse) than for the original design, but still within the range permitted by code?
- What would you have done if your insurance carrier had said to "keep quiet"?
- What would you have done if the building owner, or the city, had said to "keep quiet"?
- Who should pay for the repair?

9.4   IS IT OK FOR ME TO BE WORKING ON THIS PROJECT?

Most of the cases we have mentioned seem to be more about engineering practice than about design *per se*. But, in fact, they are about the same issues because the engineers involved were clearly working to realize and support designs they had created. On the other hand, perhaps a more telling point arises when we are asked to design a product that we think needn't be made, or even shouldn't be made. We referred in Chapter 5 to the design of a cigarette lighter, which we also thought of as an igniter of leafy matter. While this example seems trivial and even nonsensical, it points to another facet of loyalties divided. It suggests that designing cigarette lighters is, somehow, not morally correct. Certainly, in the United States today, it would seem that designing cigarette lighters and cigarette-making machinery might be viewed as at least politically incorrect, and therefore perhaps morally incorrect, by a large fraction of the population. On the other hand, isn't it up to individuals to choose to smoke or not? If our products are legal, then, shouldn't we allow ourselves to design them and, at the same time, to feel comfortable that we are designing them?

A more glaring instance along these lines arises if we look at the designers of artifacts such as the large scale ovens and their specialized buildings made in Germany in the 1930s and 1940s, and to some, the nuclear weapons designed both here and in the Soviet Union since the late 1940s. The technologies were entirely different, and while some engineers and physicists were excited by the intellectual challenge of designing devices that could harness nuclear fission, it's hard to imagine oven designers having similar feelings. So, were these two sets of designers merely being loyal to their clients? To their governments and to their societies? If so, where on the scale were their loyalties toward "human welfare and the environment"?

Here we have raised an issue that could be intensely personal, that is, should I be working on this design project? Unfortunately, the answer is, "It depends." There is no way to know in advance when a serious conflict of obligations and loyalties will arise in any of our lives. Nor can we know the specific personal and professional circumstances within which that conflict will be embedded. Nor, unfortunately, is there a single answer to which we can point. If faced with a daunting conflict, we can only hope that we are prepared by our upbringing, our maturity, and our ability to think and reflect about the issues that we have so briefly raised herein.

9.5 NOTES

Section 9.1: (Martin and Schinzinger 1996) and (Glazer and Glazer 1989) are very interesting, useful, and readable books on, respectively, engineering ethics and whistleblowing. Ethics emerges as a major theme in Harr's (1995) tale of a civil lawsuit spawned by inadequate toxic waste cleanup.

Section 9.2: The definitions of agency- and identification-loyalty are derived from (Martin and Schinzinger 1996).

Section 9.3: The Citicorp Center case is adapted from (Morgenstern 1995) and (Goldstein and Rubin 1996). We were greatly helped by William LeMessurier's review of the material.

Section 9.4: For the sake of brevity, we cite only (Arendt 1963) and (Harr 1995) among the many books about the deep and complex issues raised in this section.

9.6 EXERCISES

9.1 Is there a difference between ethics and morals?

9.2 Identify the stakeholders that the design team at HMCI must recognize as it develops its design for the portable electric guitar. Are there obligations to these stakeholders that you should consider and could they conflict with what your client has asked you to do?

9.3 As an engineer testing designs for electronic components, you discover that they fail in a particular location. Subsequent investigation shows that the failures are due to a nearby high-powered radar facility. While you can shield your own designs so that they will work in this environment, you also notice that there is an adjacent nursery school. What actions, if any, should you take?

9.4 You are considering a safety test for a newly designed device. Your supervisor instructs you not to perform this test because the relevant government regulations are silent on this aspect of the design. What actions, if any, should you take?

9.5 As a result of previous experiences as a designer of electronic packaging, you understand a sophisticated heat-treatment process that has not been patented, although it is considered company confidential. In a new job, you are designing beverage containers for BJIC and you believe that this heat treatment process could be effectively used. Can you use your prior knowledge?

9.6 With reference to Exercise 9.5, suppose your employer is a nonprofit organization that is committed to supplying food to disaster victims. Would that change the actions you might take?

9.7 You are asked to provide a reference for a member of your design team, Jim, in connection with a job application he has filed. You have not been happy with Jim's performance, but you believe that he might do better in a different setting. While you are hopeful that you can replace Jim, you also feel obligated to provide an honest appraisal of Jim's potential. What should you do?

9.8 With reference to Exercise 9.7, would your answer change if you knew that you could not replace Jim?

Chapter 10

The City Square
Transportation Hub:
A Design Case Study

Can you show me an example from engineering practice?

In this chapter we want to demonstrate that both the design framework and its associated formal design tools that we have described in the rest of this book are relevant, useful, and actually used. We will do this by describing a large, highly visible, civil engineering project in Charlestown, Massachusetts, a part of the City of Boston. The project, which unfolded over a period of almost fifty years, has, in the end, come to be known as the Central Artery North Area or CANA project. Why did we choose to make a case study of *this* project, with its long time frame, wide scope, and very large scale? Frankly, we selected the CANA project for discussion because its design process is well documented. It serves to demonstrate that the conceptual design tools are useful in billion-dollar ventures, just as they have proven applicable to designing beverage containers and chicken coops.

This project is also instructive in terms of the issues raised in Chapter 8 and, especially, Chapter 9. Because the project took place over such a long time period, the interests and concerns of a number of stakeholders were ultimately raised and heard. At the same time, the issues that engineers were expected to address as part of their professional obligations shifted to include a wider range of interests and concerns. We have used this expansion of consciousness to show just how the formal design tools can be used to document the interests of clients, users, and the affected public. One caution is in order, however. There is a natural temptation to assume that the engineers and planners working toward the end of the CANA project were somehow more ethical or responsible than were earlier ones. This is almost certainly unfair to those engineers who worked on the project in the 1950s and 1960s. Just as society and the profession have matured over the past fifty years, so too have its engineering practitioners. All of us are, to some extent, products of our times.

There are risks associated with following the historical evolution of a project. Nevertheless, that is the basic structure we follow here. First we look at one of the earliest attempts to address what was originally perceived as, basically, a transportation problem. We use objectives trees to examine the nature of the problem as it appeared to a team of engineers and planners in the 1950s and 1960s. Following this, some of the changes that occurred in the project and in the accompanying social order are briefly examined. Finally, we look at some of the design considerations that ultimately came into play in the final studies that preceded construction. Just as the perceptions and understanding of the engineers have changed over the nearly fifty years of the project, so too have the design tools that were available to the design teams. In the earliest studies, the teams used few formal design or representation methods. In those cases, we have applied the techniques described in this book to document what were usually verbal statements and assessments of objectives and constraints. Many of the formal tools were used in the later analyses, and examples from the actual notebooks of the engineers and planners are included. While not all of the design tools were used either directly or exactly as we have formulated them, it is nonetheless striking that so many were.

10.1   THE HISTORY AND EVOLUTION OF THE CITY SQUARE PROJECT

Charlestown is known to many as the town to which Paul Revere crossed by boat from Boston's North End to begin his famous ride. Administratively, however, Charlestown is one part of the City of Boston (see the map of Boston shown in Figure 10.1). It lies north of what many people consider the rest of Boston, separated from the city by the Charles River. Still further north, across the Mystic River, lie many of the bedroom suburbs that turn out to be one of the principal driving forces behind this project. Indeed, Charlestown's location has lent itself to serving as a transportation corridor between Boston and the northern communities of the Commonwealth (or state) of Massachusetts. Charlestown's residents have always identified themselves as "Townies," not as Bostonians. Like many of Boston's neighborhoods, Charlestown has much of the character of being a distinct place of its own. Immediately east of Charlestown is Boston Harbor, where the historic frigate, the U.S.S. Constitution, is maintained. Several Navy yards, commercial shipyards, and many port warehouses have long given Charlestown something of an industrial character with a strong nautical flavor.

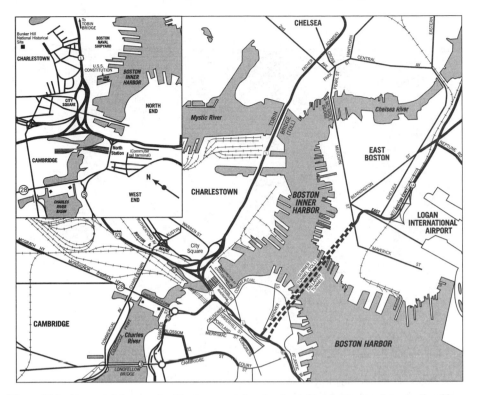

Figure 10.1 A map of the greater Boston area showing how Charlestown relates to the City of Boston. Note the transportation corridor from the north, that is, the highways coming over the Mystic River via the Tobin Bridge, and heading south through Charlestown toward the Charles River crossing into Boston itself.

10.1.1 1948–62: Engineers First Think About City Square

At the end of World War II, urban residents began to leave cities like Boston. They moved in droves to suburbs, skipping past communities such as Charlestown and moving instead into bedroom communities along Boston's North Shore. (There was also migration inland to towns and suburbs such as Concord, Lexington, and Newton, and toward communities along the South Shore, but these movements did not have a direct impact on the CANA project.) As the number of suburban residents increased, so too did the demand for highway facilities to support daily commutes into and out of Boston. To the north of the city, this demand was met in part by the construction of an expanded set of highways and their related infrastructure. For example, a major bridge, the Tobin Bridge, was constructed over the Mystic River in the late 1940s, connecting eventually into Boston's Fitzgerald Expressway from the north, giving commuters and motor carriers a direct access corridor into Boston that passed through (and effectively bypassed) Charlestown itself.

The placement of the highway connection, however, was problematic almost from its inception. As can be seen in Figure 10.2, the end of the Tobin Bridge

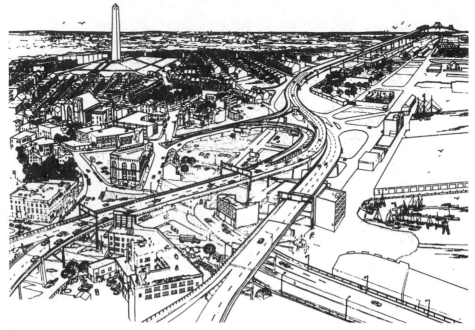

Figure 10.2 A sketch of the Charlestown City Square Area as it looked from the 1960s until the completion of the CANA Project. Notice how the community is both cut off from the waterfront and dominated by the elevated highways. The dangerously crooked off-ramp from the bridge access road is also clearly visible. The sharp left and (immediately following) right turns into the various loops is one of the most dangerous sections of the U.S. Interstate System. From (Berger 1981).

access road was, for a variety of political and other reasons, designed to have a sharp eastward turn, making the exit off the bridge difficult and dangerous. (Local lore had it that the road was located in this way in order to avoid moving a tavern belonging to a well-connected friend of influential politicians, but there seems to be no documentary evidence that supports this assertion.) The highway access from the north formed a maze of elevated highways and access roads that had the effect of cutting off Charlestown from the waterfront, and turning City Square into nothing more than a dark and noisy transportation corridor. The location of access roads into Charlestown from the highway network also had the effect of encouraging drivers who wanted to avoid Boston's highways to the south to leave the elevated highways and use local and residential streets as through roads. In the early years of the Tobin Bridge (1950–60), the Bridge Authority considered there to be too little traffic, and so it wanted engineers to redesign the roads to make the trip from the north more appealing and so generate more toll revenue. In later years, it became apparent that the road network could not handle even the existing traffic, and it became necessary to find ways to both handle traffic more efficiently and encourage commuters not to drive.

During the same period, Charlestown was undergoing a period of economic and social change. Long a neighborhood noted for social and familial stability, it began to lose young residents who moved to the suburbs, away from homes that families had lived in for several generations. As new people moved into this older housing, only the less economically able remained, and the community suffered a gradual decline in real estate values, job opportunities, and in neighborhood identity. These changes, which were common to most urban areas in the 1950s and 1960s, led government planners to consider ways to renew and revitalize urban life. The term *urban renewal* was used to identify and characterize a wide array of plans and programs designed to improve the quality of urban life, many of which had the unfortunate effect of destroying the fragile character of the neighborhoods they sought to save. In Charlestown, for example, a number of older homes and buildings, many being interesting both historically and architecturally, were torn down to make way for new apartment buildings or commercial spaces. The effect of this "renewal" was to make it less than clear that the long-term interest of the residents were being properly identified and represented, and how those interests could be best met.

Against this backdrop, the North Terminal Area Study (NTAS) was commissioned and undertaken as a coordinated effort under the sponsorship of a group of public and private agencies and organizations, including:

- the Boston and Maine Railroad,
- the Boston Redevelopment Authority,
- the City of Cambridge,
- the Massachusetts Department of Public Works,
- the Massachusetts Port Authority,
- the Mass Transportation Commission,
- the Metropolitan District Commission, and
- the Metropolitan Transit Authority.

These groups represented virtually all of the major players with transportation or land use interests at that time. The study itself was conducted by a team of two Illinois firms that specialized in transportation and land use analysis, and especially highway engineering (Barton-Aschman Associates, Inc., and Alfred Benesch and Company), and a leading expert in mass transportation planning (Colonel S.H. Bingham of New York). The firms were asked to assess the present conditions, project the future conditions, and recommend changes to the transportation network that would meet its future needs. Figure 10.3, taken from the NTAS final report, shows the factors that the study sponsors felt demonstrated the need for action. These factors included traffic congestion at various points, the planned relocation of transit facilities, urban renewal and reconstruction programs, and the planned construction of a number of highway projects.

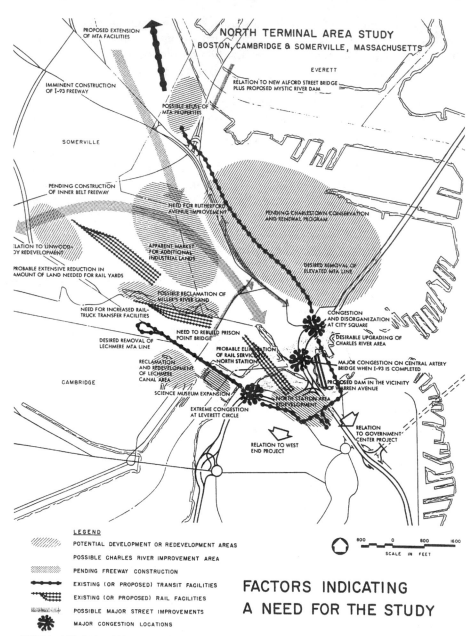

Figure 10.3 This figure is taken from the NTAS study and highlights what were perceived to be the key issues and concerns at the time that study was done. The graphical method of presenting data shown here was widely used at the time of the study and is still in use today.

The team then decomposed or subdivided the problem into three main issues: land use considerations, highway needs, and mass transportation needs. Specifically, the team looked at five specific subtasks, that is, the design team:

1. compiled objectives that related transportation to land use, and then developed conceptual, generalized plans;
2. developed a comprehensive street and highway plan that recognized both traffic needs and land development;
3. analyzed existing and developed future traffic flow patterns and made assignments to local streets, highways, and mass transit facilities;
4. determined the engineering feasibility and preliminary cost estimates for proposed solutions; and
5. developed a mass transportation plan that included general design, engineering feasibility, and preliminary cost estimates.

It is beyond our scope and our current interest to examine each of these tasks in detail. What is of particular concern to us is the first of these, compiling the objectives and developing the conceptual, generalized plans that were reported in the NTAS. We will explore these in more detail in the following sections.

10.1.1.1   The Objectives of the North Terminal Area Study

The North Terminal Area Study (NTAS) team identified, essentially, three major objectives and a series of subordinate objectives or subobjectives that an effective transportation system would have to meet. These objectives are shown in Figure 10.4 in the form that they were presented in the actual NTAS report. In Figure 10.5 we present the same objectives in an objectives tree based on that NTAS report, but developed by the authors. The initial graphic of Figure 10.4, while visually appealing, makes it difficult to relate the objectives to each other and to the overall project. Interestingly, the primary focus of the study, insuring adequate capacity for the hundreds of thousands of daily commuters through the study area, is considered so obvious that it is not even expressly entered into the figure.

Several points should be made concerning the objectives given in the NTAS report and highlighted in our objectives tree. First, it is clear that the engineers and planners understood a great deal about the complexity of the urban environment in which they were operating. They were aware of the need for better waterfront access, understood that the neighborhoods could be adversely affected by traffic and should be protected from it, and saw the need to coordinate their efforts with other major projects already underway. The NTAS team also clearly wanted to locate their transportation solutions in ways that did as little harm to the community as possible. For example, the transit station was to be located at a site that would promote commercial development, without inducing "neighborhood blight," which at that time was considered to be an outcome of mass transit.

The team's report offered no direct evidence that it ever organized its objectives in a manner as coherent as an objectives tree, or that it formally weighted its objectives or attached formal metrics to them. Nonetheless, it is clear that there were specific weights and values, and that these shaped both the alternatives considered and the NTAS team's specific recommendations.

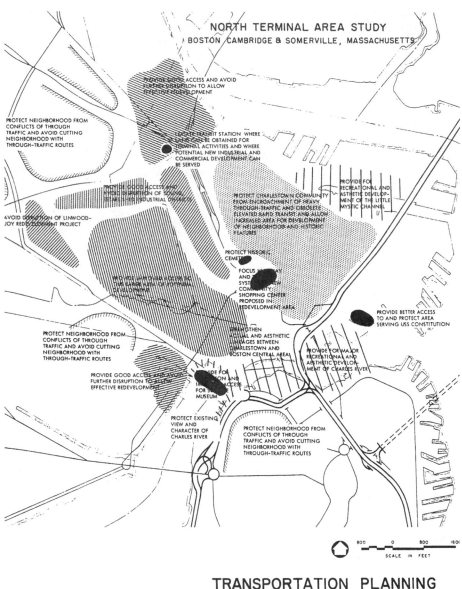

**TRANSPORTATION PLANNING
CONSIDERATIONS**

Figure 10.4 This figure, also from the NTAS report, shows some of the objectives and concerns (labeled considerations in the report) that the NTAS team found. It is instructive to compare this presentation with Figure 10.5 in terms of how easy it is to use to trace the relationships among the objectives.

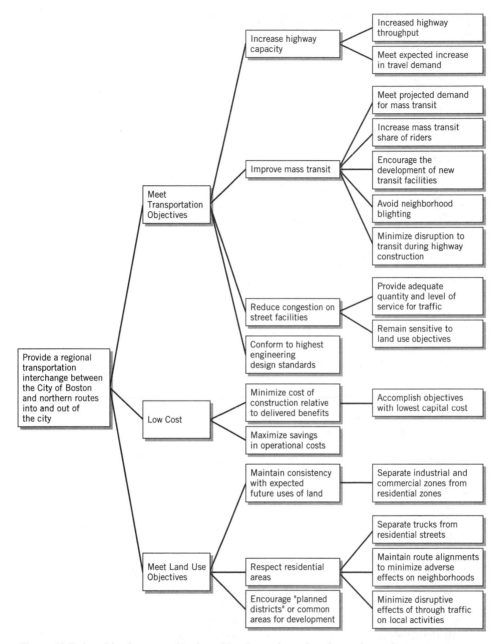

Figure 10.5 An objectives tree developed by the authors that shows the various concerns raised in the NTAS report. While no weights were ever formally given in the report, the context makes it easy to see that the highway objectives dominated both the analysis and the report's recommendations.

10.1.1.2   *The Methods Applied by the NTAS Design Team*

The basic process of the NTAS team was to list its concerns in each area (land use, traffic, and mass transit), use these concerns to develop some general design approaches, and then consider various parametric designs in the light of these concerns and approaches. That is, the team never really opened its design space very widely, so its alternatives were essentially modifications of the existing highway and street network in order to increase traffic throughput. We can see evidence of this in their final report. After laying out and quantifying the various considerations, such as those in Figure 10.4, the team turned almost immediately to particular geometric design alternatives, which in civil engineering terms means configuring highways, roads, streets, and associated structures to meet traffic and other demands. Figure 10.6, for example, shows one of a number of renderings of possible structural alternatives for expanding the elevated highway. From this point on, the concern of the report was to identify the roads that needed to be widened, those that needed to be relocated, and those parts of the neighborhood that had to be sacrificed to make the roadwork possible.

The alternatives considered thus included:

- restricting all nonlocal traffic to highways;
- expanding the capacity of the existing set of highways, ramps, and local arterials;
- expanding the expressway system to increase its capacity;
- upgrading surface routes, including adding some new bridges and local roads; and
- making greater use of other roads to remove traffic from the City Square area and divert it to alternate paths into Boston.

Again, the details of the alternatives themselves are not very interesting to us, not only because none of them were ultimately adopted, but also because our focus is not on highway design. However, what is striking about these alternatives is that all of them are essentially working within the same, limited design space. All of them are basically traffic flow alternatives. While it seems only natural to wonder whether or not a wider set of alternatives exists, the team apparently gave little or no consideration to finding ways to restrict the growth of traffic into and out of the city, perhaps through means such as tolls. Similarly, transit alternatives were considered primarily as means of meeting existing and projected demand, with little or no consideration being given to exploring how that demand might be shifted. Finally, the effects of the design alternatives on the community of Charlestown seem to have been considered only in terms of whether or not they made existing conditions worse. There appears to have been little or no consideration given to expanding the design space to consider what, if anything, could be done to directly affect the changes in the community in a positive manner.

The evaluation of the alternatives appears to have been based on two simple metrics—traffic capacity and cost. Only alternatives that were projected to meet the traffic demands were given serious consideration. Among this subset of

NORTH TERMINAL AREA STUDY
BOSTON, CAMBRIDGE & SOMERVILLE, MASSACHUSETTS

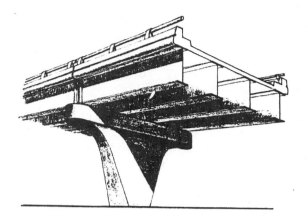

RAMP PIER

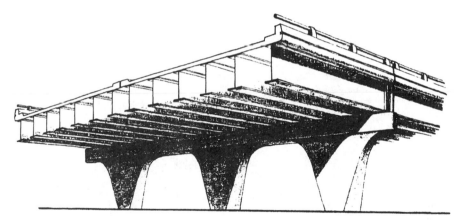

TYPICAL PIER

Figure 10.6 This figure, showing sketches for pier designs, is taken from the NTAS Final Report. It shows how far removed the team was from conceptual design by the end of the study. This change of focus occurred without anyone—on the design team or not—ever really asking what would be the implications of limiting the conceptual design space.

design alternatives, the single deciding factor appears to have been the construction cost. It is interesting to ask, with the benefit of hindsight, whether or not this was appropriate even for the limited set of alternatives considered. Recall, for example, the set of "considerations" in Figure 10.4, or the objectives that are collected and organized in Figure 10.5. These suggest a number of other measures that could well have been applied.

Another important question that the above process raises is, Are there ethical dimensions to deciding to limit a design space, assuming a design team has the freedom to consider alternative concepts? It is certainly not our intent to undermine or even criticize the engineers who worked on the project *at that time*. However, in today's world, an engineer who was faced with a set of choices and information would clearly be called upon to consider many more factors than just cost and traffic, and to enlarge the design space to ensure that those concerns are raised.

10.1.1.3 The Results of the NTAS

Figure 10.7 shows the specific design recommendation that the NTAS design team proposed. We see that the proposed design adds several additional roads and expressways into the City Square area, and it enhances traffic flow on the highway system. However, the proposed system does nothing to improve the access of the Charlestown community to the waterfront, while it essentially exacerbates City Square's role as a transportation corridor. Interestingly, the proposed design by the NTAS team leaves intact the eastward jog in the exit ramp, although the ramp's curve is extended to help reduce the highway safety problem.

The proposed solution was not adopted, but it did provide a basis for further discussions among transportation engineers, land use planners, *and* the communities affected by the road system.

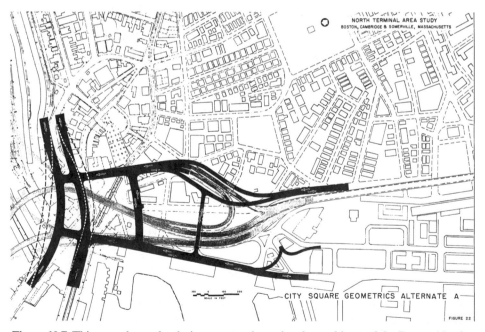

Figure 10.7 This map shows the design proposed to solve the problems of the Boston North Terminal Area. Note that the solution proposed increasing the number and size of expressways passing over City Square.

10.1.2 1962–74: Engineers Rethink City Square

During the 1960s and into the 1970s, City Square continued to be a topic of interest for engineers and planners. Several studies proposed various new highways or bridges into Boston, all placed above City Square. In each of these studies, the dominant concern was to find ways to manage the increasing traffic levels, while addressing at the same time the strange geometry of the roadway at the end of the Tobin Bridge. One alternative, for example, suggested a multilevel bridge into Boston that would raise the already elevated Fitzgerald Expressway still higher and connect it directly into the east-west highways going into Boston. Other, less dramatic proposals called for realigning the Fitzgerald Expressway and building a series of smaller-scale bridges into the rest of the city.

The 1960s and early 1970s were also periods of major social and political change in the United States. The country was torn by a series of assassinations, the continuation and at least partial success of the civil rights movement, and a near uprising in response to an unpopular war in Southeast Asia. Campus and community activism began to focus not only on the government's conduct of foreign policy, but also on its role in community development and preservation. One very profound change was an emerging concern with the environment. The environmental movement was able not only to raise the consciousness of individuals, but it also gained passage of legislation that required that the environmental impacts of major public works be examined, documented, and seriously considered when engineering decisions were being made. This legislation, initially just at the Federal level, mandated that future public works projects could not go forward unless all of the possible environmental effects of the project were identified and examined and an Environmental Impact Statement (EIS) was written and published.

Needless to say, engineers were affected by these social changes. In addition to meeting formal, legal requirements that called for a range of impacts to be considered in their projects, many engineers embraced community activism as an answer to defining who were their clients and stakeholders. In short, new public reporting requirements gave engineers access to citizens who were affected by the projects, even if these stakeholders seemingly had little or no direct decision making authority.

In concert with these large-scale social and political changes during this era, there were also some decisions made regarding the roads that connected to the City Square highways and local streets. The proposed Inner Belt, which was to loop around the city from relatively close in, was canceled because politicians and engineers realized that the cost of such a road to the social fabric was simply too high. Serious restrictions on the growth of new highways and the expansion of old ones were also put into place, primarily in the form of executive decisions by the governor of Massachusetts. At the same time, urban planners came to realize that parking and local traffic in Boston had reached the point that new roads into the city might do more harm than good.

It was in this climate in the early 1970s that engineers began to look at the environmental impacts of a series of alternatives that had been offered by various agencies and groups during the previous decade.

10.1.2.1 The Objectives of the New North Terminal Study

While an environmental impact study is not generally the same as a conceptual design project of the sort we have discussed throughout this book, in this particular case they actually share many common features. A conceptual design exercise has at its core the identification and clarification of the core problems, the generation and analysis of alternatives, and, hopefully, the selection of a concept that, if properly designed and further developed, will lead to a successful outcome for all of the key stakeholders. This was, in many respects, the intent of the EIS study commissioned by the Massachusetts Department of Public Works in 1972. After highlighting the various groups in conflict and some of the alternatives that had been proposed, a draft EIS stated that its purposes were "to help resolve conflicting viewpoints as to the proper solution for the area" and to provide "a means of delineating the major factors of which a rational choice solution could be made." Interestingly, the study behind the EIS had the effect of galvanizing some of the stakeholders and leading to significant revision of the alternatives that were considered.

10.1.2.2 The Methods and Approaches as the Study Developed

As participants in a study examining the environmental impacts of a number of alternatives, the engineers were required by law to present the alternatives and their methods for studying them in a number of public forums. While this process was primarily directed toward the public presentation of the design team's activities and findings, it effectively required the team to fomally consider how it generated, evaluated, and selected among design alternatives. (This topic has been one of our continuing themes, especially in Chapter 6.) In addition, this process requirement virtually guaranteed that new stakeholders would be identified and given the opportunity to express their viewpoints. The scope and magnitude of the proposed alternatives were such that public hearings were held throughout Boston. At the hearings in Charlestown, a great deal of outrage was expressed at the proposed taking of homes and residential property under each of the major alternatives and at the apparent disregard for any specific benefits that might accrue to the local communities. A particularly strong objection was raised to alternatives that called for Charlestown to host a series of "park and ride" locations. These alternatives would allow commuters heading for Boston to drive as far as Charlestown and would encourage them to park there and continue into the city using mass transit. The result for Charlestown would be the continuation of an unappealing status quo, save for the addition of a series of parking lots to an already uninviting area!

As a consequence of the several public meetings held in the greater Boston area, a number of new concerns were raised and a method for assessing design alternatives was developed. Figure 10.8 shows an instructive summary of the impacts by alternative. The table therein lists several alternatives, identified by letter, and the number of potential benefits or adverse effects associated with each alternative. Except for construction cost, every alternative has been assessed in terms of whether the impacts are positive (recorded as plusses), negative (minuses), or neutral (zeroes). For our purposes, the particular values assigned by the study team to specific alternatives really don't matter very much, nor are we especially interested in what the particular alternatives were. What is noteworthy is that this table shows, albeit indirectly, which objectives were of interest to study participants, it hints at the metrics applied to those objectives, and it shows how metrics were weighted here (see also Figure 6.6).

In Figure 10.9 we have taken the objectives implicit in Figure 10.8, along with some of the materials given in the textual portions of the Draft EIS, and have constructed an objectives tree. It is interesting to compare this objectives tree with that shown in Figure 10.4. Many of the issues regarding land use and planning are the same and evident in both, but the new legal requirements have clearly increased the importance of objectives that are not related directly to traffic throughput or construction cost. Indeed, the later tree (Figure 10.8) is less concerned with higher traffic throughput than it is with the "quality of the ride" for existing travelers, that is, for the users of the system.

In spite of the general, process-oriented approach we are taking to describing the City Square project, a few remarks are in order on the specific alternatives under consideration. Alternatives A and B called either for maintaining the status quo or for minimal construction that would address traffic needs that were well understood and agreed upon. Alternative C called for building a new

Figure 10.8 This chart, taken from the draft EIS for the new north terminal study, shows some of the considerations and values that were applied as various alternatives for an elevated highway over City Square were examined. Note that the number of "+" and "−" signs reflects the design team's scaling of the impacts.

Figure 1–1: Section 1.1 Summary Comparison of Impacts By Alternative						
	A	B*	C	D	E_1	E_2
1. Construction Cost (millions) (1973 dollars)	$0	$8.5	$21.1	$20.9	$20.1	$20.6
2. Traffic Benefits: Local	− −	+	+	+	+	+
3. Traffic Benefits: Expressway	0	+	+	++	+	++
4. Displacement: Housing	0	0	− − −	0	−	− −
5. Displacement: Businesses	0	0	− −	− −	− −	− − −
6. Effect on T and RR	0	0	0	−	− −	− −
7. Section 4f Lands	0	0	− − −	− −	− −	− −
8. Regional & Community Growth	0	−	−	0	− −	− − −
9. Public Facilities	0	0	− − −	− −	− −	− −
10. Aesthetic Values	−	−	− − −	− −	− − −	− −
11. Noise	−	−	−	−	−	−
12. Air Pollution (1 hour and 8 hour standards)						
1980 (standards exceeded)	− −	−	−	− −	− −	N.A.
2000 (compliance)	+	++	+	+	+	N.A.
13. Community Cohesion	− −	−	− − −	−	−	− − −
14. Conservation & Preservation	0	−	− −	−	−	−

*Alternatives B_1 and B_2 are not shown because they are not evaluated in detail.

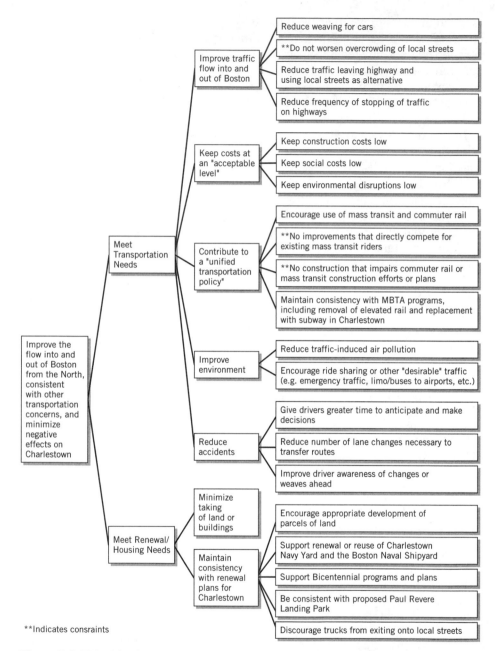

**Indicates consraints

Figure 10.9 This objectives tree was developed by the authors to reflect the draft EIS for the new north terminal study. Note that the focus on the highway users has begun to shift from simple throughput, and that some other land use and environmental concerns are now being acknowledged more openly.

bridge, across the Charles River, directly from the elevated highway in City Square to Boston. It would result in thirty residential and four business properties being taken, as well as in numerous construction-related disruptions. Virtually all of the adverse effects would occur in City Square. This alternative was vehemently opposed by Charletown residents at public hearings. Alternatives D and E, offered in several versions, called for taking various numbers of residential or business properties, ranging from eight to twelve, and involved continuing the elevated highways through City Square. One of the engineers who participated in the public hearings and later became Massachusetts Secretary of Transportation, Fred Salvucci, said that when these alternatives were proposed, he expected the residents to be gratified that so few homes would be taken. In fact, however, they were outraged that the best effort engineers and planners could produce would not actually improve their community, but would only limit the damage. Salvucci indicated that this was one of the most significant moments of his professional life as he came to realize that along with his choice of career had come obligations to others.

10.1.2.3 The Results of North Terminal EIS Study

While the EIS described above did not result in a particular alternative being selected and adopted, it had a profound effect on the final outcome—and for several reasons. The study took place at the same time as another major construction project whose planning had just begun. That project dealt with the redesign of the Central Artery, an elevated highway across downtown Boston that essentially cut the entire city off from its waterfront. A plan had been proposed to relocate the Central Artery underground, in a tunnel, which would allow Boston to reclaim its access to the waterfront and, simultaneously, create a new urban space. Depressing most of the entire Central Artery below Boston also forced a reconsideration of how the city's entire landscape was being adversely affected by elevated highways. The North Terminal EIS study provided valuable insights into how such projects could either improve neighborhoods or worsen them. Perhaps more importantly, the process led Charlestown community activists to consider how City Square could reunite their community both with its own waterfront and the role that waterfront had played in Charlestown's traditions. The project also enabled communication between the community activists and the engineers and planners who were designing the new Central Artery project.

10.1.3 1974–81: "An honorable place to live . . ."

This period's outcome actually began with some rumblings in October 1973. Alan Altshuler, then Massachusetts Secretary of Transportation, responding to public input, reached several conclusions about how the area to the north of Boston should be treated. Altshuler turned those conclusions into constraints, among which were the strictures that no houses in Charlestown would be taken for highway projects and no decisions would be made about the northern area until a recently initiated study of the Central Artery was substantially complete. The proposal to move the Central Artery into a tunnel revitalized in many ways the thinking of highway engineers and planners in the region as they considered how they might construct alternative routes that were more consistent with the city as a space for *living*, as well as for working.

One of the prime beneficiaries of this revised approach was Charlestown, whose residents continued to clamor for a better alternative for City Square. During the mid-1970s, community groups were able to seize upon the Central Artery and argue that a similar alternative should be available to them. By 1978 the community had convinced the engineering staffs of the city of Boston and the Commonwealth of Massachusetts that they would successfully oppose any alternative that did not include two elements: depressing the highway from the Tobin Bridge under City Square and removing the twisted jog in the bridge's exit ramp. Since Federal money for planning and constructing the Central Artery had just become available, this alternative, which came to be known as the Central Artery North Area (CANA) project, was readily accepted by all of the parties involved. Thus, by 1978 the decision to depress the arteries had, essentially, been reached.

Now, it might seem that the decision to build a tunnel under City Square effectively ended conceptual design work on this project. In fact, it merely changed the basis for conceptual design and moved other aspects of the project into preliminary design. As we have said before, design is by nature an iterative process in which designers and engineers move repeatedly through stages (conceptual design, preliminary design, and lastly, detailed design), often returning to an earlier stage for some part of, or even all of, a project. In this case, the conclusion that the best conceptual alternative was to depress a major highway under an urban area created several new conceptual design issues. In particular, since the older elevated highway was to be removed, twelve acres of land that had not been part of the cityscape would become available. This, in turn, required conceptual design for an appropriate new network of streets. Similarly, the new roadway under City Square would have to be connected to the rest of the highway network upon returning to ground level on the other side of the square. Finally, the preliminary design of the tunnel itself would require major decisions on how a tunnel concept would be implemented.

It is instructive to examine here how the engineers and planners who were given the task of developing a conceptual design for the new street network went about their business. They developed a set of twenty-two conceptual designs, ranging from doing essentially nothing (i.e., constructing the roads to continue more or less as they had before, when the elevated highway was running over Charlestown) to selecting various ways for traffic to exit the now-underground highway and flow into city streets. Of these twenty-two alternatives, five were eliminated either because they were essentially the same as others or because they were impractical. Four more were also removed from further consideration because they violated constraints that had been agreed to by the stakeholders. Thirteen alternatives remained, and they were subjected to a preliminary analysis in terms of traffic flow, urban design, and community development effects.

Eight of these alternatives were then chosen for further analysis, including two that would have been rejected but for the wishes of community representatives. These eight alternatives were subjected to further, detailed traffic analyses, and the top four were taken to the community so that it could choose. Once again the community asked that one of the previously rejected alternatives be added back into the group under consideration. Two of the (remaining) top four alternatives were subsequently rejected when they were found to have adverse impacts either on pedestrians or on land development.

At this stage, the designers returned to the drawing board and refined their designs for the surviving alternatives, producing some ten new modifications of the potentially acceptable designs. The numerical metrics that were applied to these alternatives led to a preference for one particular scheme. As it happened, the community and the City of Boston did not like that particular scheme because of its impact on local traffic and pedestrian circulation. Therefore, still another alternative was adopted and subsequently built.

The pattern behind this unfolding design process had a number of features that were simply not seen in the earlier design studies. Perhaps the most striking of these was the expansion of the stakeholder set to include the Charlestown community as an equal both with the public agencies who would oversee—and pay for—the project and with the highway users who would be the most obvious consumers of the design. A second feature was the implementation of the process as a highly iterative design activity. Note that the designers never "fell in love" with a particular design alternative. They could therefore modify and enhance alternatives in order to better meet their clients' needs. Another feature of this process was that the metrics defining a good design changed in significant ways; for example, traffic circulation and land development were now seen as important aspects of a successful design. Even more striking in this regard, however, was the way that the weights applied to these measures changed from those used in the early 1960s. Since the cost of the project was to be borne largely by the Federal budget, rather than being paid for with local monies, the cost of construction was no longer *the* dominant factor. On the other hand, pedestrian traffic and local traffic were now of great significance. These two issues could have been measured in the earlier study, but they were not of sufficient weight to really come into play as decisions were being made.

In Figure 10.10 we show an objectives tree that was developed by the authors to show some of the considerations in the 1970s studies. Because the stated concerns had been widened, a number of areas are not fully shown in the tree. It is obvious, however, that many factors that had not given much significance previously have here become important objectives. For example, the impact of noise and dust on the psychology of City Square appears here for the first time.

Over the same period of time that this conceptual design process was going on, the engineers were also considering alternative ways of building the tunnel itself and of linking its underground road to the expressway network. In the next section we will present examples of design tools that were used by the design team as a way of demonstrating how some of the methods described in Chapters 3–7 were used in a major design project.

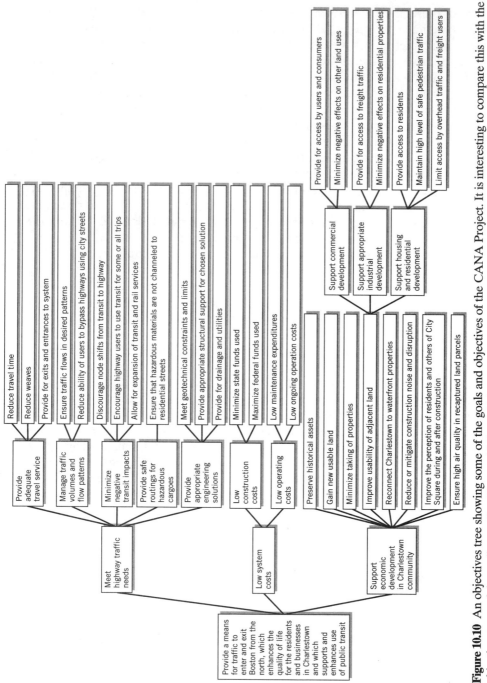

Figure 10.10 An objectives tree showing some of the goals and objectives of the CANA Project. It is interesting to compare this with the objectives trees of the earlier studies, Figures 10.5 and 10.9.

10.2 FORMAL DESIGN METHODS LEADING TO A SUCCESSFUL OUTCOME

Having set some of the context in which the City Square project took place, we return to our main focus, the application of design processes and tools in a "real world" setting. In the previous section the present authors took the liberty of drawing objectives trees in order to highlight the objectives and constraints that the various design teams faced. This was an inevitable consequence of our wanting to show the changes in objectives—and implicitly, the growing set of included stakeholders—in several studies over several decades. In this section, however, we will look primarily at the tools and methods used by one of the design teams, namely those working on the last project, the CANA project.

For the sake of completeness, our presentation follows the general pattern of formal conceptual design that we have articulated. But the emergence of this pattern is more than an affectation on our part. Put simply, it represents in a very real way the processes that the designers themselves have documented.

10.2.1 Objectives Elicited and Articulated

We have already discussed the objectives in Section 10.1 and shown how the goals and objectives of the engineering design teams came to change over the life of the project. This reflected both new perspectives of who the client(s) and other stakeholders were and changing perspectives on the role of the engineer in projects that have enormous social and economic implications.

Earlier in this chapter we showed objectives trees for each of the design studies (Figures 10.4, 10.5, and 10.10). The most striking element of these is the emergence of a wider range of concerns as time progressed. This is not surprising, but it should serve to highlight the need for engineers and designers to remain sensitive to the context in which a design will be used. No one can be expected to accurately predict the future course of the society in which design will take place, but engineers may have a particular obligation to consider the future and the impacts on the lives of stakeholders when designing products that are either long-lived, such as highways, or very widespread in their usage, such as successful consumer products.

10.2.2 Functions that Realize Articulated Objectives

The CANA project serves as an excellent illustration of the importance of understanding functions. In fact, the engineers and planners who conducted the project did exactly the sort of functional analysis detailed in Chapter 5. Recall that functions are the verb-object combinations that describe the things which an artifact must *do*. These functions can be broken down into basic functions and secondary functions. Basic functions are those that the artifact must perform in order to accomplish its fundamental mission, while secondary functions are those that the artifact must do as a result of other functions. We suggested that the engineer should use the most general language possible for functions during the conceptual design phase in order to avoid limiting the design space needlessly.

Figures 10.11 and 10.12 display the functions for a bridge and for a tunnel, respectively. The figures are part of the notes of the engineering team that performed what they called the *value analysis* of various alternatives. Bridges and tunnels were the two most basic alternatives that were considered by the design team during the conceptual design phase. Both alternatives have *connect points* and *support weight* as their basic function. Both alternatives also have several secondary functions, including *resist forces* and *communicate with drivers*.

Note that the design team essentially combined enumeration and dissection, that is, they first considered known engineering solutions to river crossing and then determined the general function as a means of gaining leverage over basic functions. The team extended this approach after the decision was made to use a tunnel under City Square when they analyzed ways to implement the tunnel and to connect that tunnel to the remainder of the road network (Figure 10.13).

10.2.3 Metrics that Measure Articulated Objectives

Having determined and articulated the objectives of the CANA, the design team needed a means to assess how well each design alternative might meet those objectives. This is the role of metrics, which we discussed in Section 5.3. Metrics are methods of measuring how well an objective is realized by a particular alternative. Recall that a good metric has a number of qualities, including having an appropriate level of accuracy, being cost effective, and being readily understood by users and stakeholders. Table 10.1 shows a listing of some of the metrics that were used by the CANA design team. (A full listing would be beyond our present scope and, likely, the interest levels of most readers.)

It is worth noting that several of the metrics derive from the kinds of activities that most engineers would find comfortable, such as assessing performance either by exercising a computer model (such as the CALINE3 model that relates traffic conditions to resulting pollutant levels) under specified conditions or by applying a standard (such as the Transportation Research Board Circular #212 that defines various levels of traffic service) for a particular level of service. In other cases, however, the metrics include judgment calls made by the design team, such as deciding to take the number of unusable parcels of real estate as a measure of good urban design or to estimate the ability of a network to accommodate future pedestrian traffic as a measure of traffic service levels. In such cases the team must clearly decide how small or oddly shaped a parcel must be to be considered unusable. Similarly, people of good will can honestly disagree about future pedestrian or bicyclist volumes in an urban area. It is incumbent upon the designers to be as explicit as possible about the underlying assumptions made and the techniques used to support them when such decisions are being made.

In other cases, not listed in the table, the design team chose to use very qualitative, highly descriptive metrics to document the expected effects of their design alternatives. For example, in the area of land and housing impacts, the team could have attempted to estimate the loss in real estate values associated with construction noise. Instead, they opted to describe the specific properties that would be impacted by noise, dust, or other undesirable side effects during

Function Analysis

Project: CENTRAL ARTERY-NORTH · BRIDGE
Item:
Date:

Quantity	Unit	Component	Function Verb	Noun	Kind	Explanation	Original Cost	Worth
		BRIDGE	CONNECT	POINTS	BASIC			
			SUPPORT	WEIGHT	BASIC			
			ACCOMODATE	MASS	BASIC			
			RESIST	FORCES	SEC.			
			ALERT	DRIVERS	SEC.			
			ALERT	PEDESTNS	SEC.			
			COMMUNICATE	DIRECTN	SEC.			
			COMMUNICATE	SPEED	SEC.			
			COMMUNICATE	CHANGE IN SPEED	SEC.			

Basic Function

AA-6

Worksheet 4

Figure 10.11 This is a functional analysis for a bridge as an alternative for moving traffic through Charlestown into Boston. It uses a combination of enumeration and dissection. Notice that the engineer has distinguished between basic and secondary functions. We have changed the engineer's handwriting to a more uniform scripted font.

Function Analysis

CENTRAL ARTERY-NORTH

Project	Item			Basic Function		Date	
TUNNEL							
Quantity	Unit	Component	Function Verb	Noun	Kind	Explanation	Original Cost	Worth

Quantity	Unit	Component	Verb	Noun	Kind	Explanation	Original Cost	Worth
		TUNNEL	CONNECT	POINTS				
			RESIST	FORCES				
			HOLD BACK	LIQUIDS				
			REGULATE	TRAFFIC				
			SUPPORT	WEIGHT				
			COMMUNICATE	DIRECTION				
			COMMUNICATE	SPEED				
			COMMUNICATE	CHANGES TO SPEED				
			ACCOMODATE	MAINT.				
			EXCHANGE	AIR				
			LIGHT	INTERIOR				
			ACCOMODATE	MASS				
		PORTAL	IDENTIFY	EXIT/ENTRANCE				
			LINK	BOAT SECTION				
			RESIST	FORCES				
		PAVING	PROVIDE	TRACTION				
			SUPPORT	WEIGHT				
			TRANSMIT	FORCES				
			ACCOMODATE	MAINT.				

AA-7

Worksheet 4

Figure 10.12 This is part of the functional analysis for a tunnel developed by one of the engineers in the CANA project. The designer has distinguished between the functions of the tunnel itself, the portal or entranceway, and the paving or road surface. Note that the terms are kept general to allow for a larger design space.

Function Analysis

CENTRAL ARTERY-NORTH

Project L-N PORTION OF LOOP   Item _____   Basic Function CONNECT POINTS   Date _____

Quantity	Unit	Component	Function Verb	Noun	Kind	Explanation	Original Cost	Worth
		LOOP (T-N PORTION)	CONNECT	POINTS	B	CONNECT POINTS	8,000,000	$400,000
		(L-N PORTION)	REGULATE	TRAFFIC	B			
		ROADWAY	CONNECT	POINTS	B			
			SUPPORT	WEIGHT	S			
			ACCOMODATE	MASS	S			
			REGULATE	FLOW	S			
			ALERT	DRIVERS	S			
			DIRECT	FLOW	S			
			DIRECT	DECLE-ERATION	S			
			DIRECT	ACCEL-ERATION	S			
			PROVIDE	REENTRY	S			
			RESIT	WEAR	S			
		RAMPS	CLEAR	OBSTACLES	B			
			SUPPORT	WEIGHT	S			
			RESIST	WEAR	S			
			RESIST	WIND	S			
		TUNNEL	HOLD BACK	LIQUIDS	B			
			CONNECT	POINTS	B			
			ACCOMODATE	TRAFFIC	B			
			EXCHANGE	AIR	S			

B1-2

209
Worksheet 4

Figure 10.13 This is part of the later functional analysis as the team considered connections between a tunnel and the rest of the roadway network. Note that the basic function, *Connect Points*, is highlighted at the top of the page.

Table 10.1 Some of the metrics used by the CANA team in deciding among the alternatives for the street network. The decision to adopt a tunnel under City Square opened a new set of conceptual design issues, including how the resultant street network should look.

Objective	Metric/Technique for Determining
Maintain Adequate Traffic Service Levels:	
Traffic Operational Characteristics	Number of undesirable traffic movements (e.g., number of left turns at high-volume intersections)
Intersection Level of Service	Traffic assignment models, then evaluation against standard, using Transportation Research Board Circular #212
Pedestrian Analysis	Capacity of network and signal systems needed to accommodate current and predicted pedestrian traffic levels
Provide for Good Urban Design	Number of unusable parcels; area (sq ft) of reclaimed usable land
Maintain Acceptable Air Quality	Predicted concentrations of carbon monoxide (CO) found using CALINE3 model together with predicted traffic counts and flow patterns under worst case meteorologic conditions
Annual Cost Comparison	Expected cost of maintenance and repair of road and associated systems
Construction Costs	Expected total cost of construction

construction. This allowed the parties involved to act in their own interests in seeking remediation while at the same time allowing community leaders and city planners to understand the effects in concrete terms. It is difficult to make a case that an abstract or highly quantitative metric would have been as understandable to the affected parties.

10.2.4 Alternative Means for Realizing the Functions

It does little good to develop and document functions unless alternatives exist to realize them. The design team developed a number of alternatives over the course of the CANA project. Consider, for example, the need for new street plans that would inevitably result from the decision to connect the Tobin Bridge to the city by a tunnel under City Square. At the conceptual design stage, for example, the team considered some twenty-two general alternatives. In the preliminary design stage the team began to consider more explicitly detailed alternatives. Similarly, the team considered several alternative tunnel types while working to depress the roadway under City Square.

In arriving at these alternatives, there is evidence that the design team used some of the brainstorming and other creative techniques discussed in Section 6.2. Consider Table 10.2, which shows the results of a team's ideas for how a tun-

Table 10.2 This is a form used by the engineers in the CANA project to document their process of generating ideas. Note that the team is encouraged to defer evaluation of the ideas until a later time.

Tunnels	Connect Points	
Study Title	Basic Function	Team

This is the creative stage of the Value Study. Generate as many ideas, processes or methods to fulfill the basic functions that the item under study must perform. Do not evaluate the ideas during this phase.

1. Rigid Frame
2. Simple Frame
3. Circular Section (Large Pipe)
4. Elliptical
5. Arch
6. Heavy Corrugated Steel
7. Pre-Cast Culverts
8. Post Tension
9. Unlined Section (As in Rock)
10. Slurry Wall
11. Driven Tunnel
12. No Tunnel
13.
14.
15.
16.
17.
18.
19.
20.
21.
22.
23.
24.
25.

List Everything—Judge Later

nel might be constructed. The alternatives include a range of engineering solutions and geometric designs. Several points about the form used (Worksheet 7) are worth noting. This worksheet focuses on means that can be used to accomplish a basic function, *connect points,* and recognizes that many teams may be looking at the same issues in a large study. The team is also cautioned, at both the top and the bottom of the form, to direct itself toward generating—not evaluating—ideas. Notice, for example, the admonition "List Everything—Judge Later" prominently placed on the bottom of the form.

We may also note that the form has much the same flavor as a single row of a morphological chart such as those discussed in Section 6.3. Given a set of such completed forms (i.e., for the tunnel, for the portals, and for the roadway connections), a designer would have the ingredients of a simple morph chart.

10.2.5 Evaluating Alternatives and Choosing A Design

Ultimately one design alternative must be chosen. The basis for this choice is, hopefully, a reasonable and rational process that addresses the strengths and weaknesses of the alternatives in light of the project's objectives. For large public projects such as the CANA project, the process of evaluating alternatives and choosing a design has to be both formal and carefully documented. In order to meet these process needs, the CANA design team took the alternatives generated by the methods outlined in Sections 10.2.1–10.2.4 and, after further analysis and consideration, applied to them a weighted scoring system much like that discussed in Section 6.4.

Figure 10.14 highlights some of the processes used to evaluate alternatives in the CANA project. Each column represents an area in which the alternatives are to be compared, including the functional performance, safety, ventilation, and maintainability of a tunnel. Each of the various criteria is given a weight, which appears to range from 4 to 16. (The basis for the weights is not clearly spelled out in the document.) Each alternative is then listed in the left-most column and scored in each category with a number of points between 0 and 100. The product of the scores and the weights are summed in the right-most column.

We can make several observations about the results shown in Figure 10.14. Note first that several of the alternatives are not completed. These alternatives have failed in some way: They failed to meet a minimum threshold value for the metric or they violated a constraint. This corresponds to our suggestion that *all* alternatives should be assessed first against constraints; this allows a team to forego further, wasted work on unacceptable alternatives. Second, note that the alternatives have already undergone some form of "pre-screening." Note the comment on the printed part of the worksheet that implies that a set of "ranking and comparison techniques" has already been applied. This reinforces our advice not to blindly apply this weighting technique, but rather to use it in support of experienced engineering judgment.

It is also important to note that the designers in this project did not depend on a rigid adherence to the numerical results. It is rather ironic that the highest scoring alternative was "No Tunnel" since it would not need any of the various

TUNNEL
Analytical Phase
CONNECT POINTS
Basic Function

List the best ideas from ranking and comparison techniques. Determine which one stacks up the best against the desired criteria.

	Desired Criteria	LOCATION	FUNC. PERFORMANCE	SAFETY/VENT.	CONSTRUCTION	STRUCTURE CONSIDERATION	MAINT.	COST	Total
		a	b	c	d	e	f	g	Total
50-9C		16	13	14	4	12	8	DIRECT	
1. PRESENT WAY		80 / 1280	90 / 1170	90 / 1260	90 / 360	90 / 1080	90 / 720	11250 LF	5870
2. SIMPLE FRAME		80 / 1280	90 / 1170	90 / 1260	90 / 360	70 / 840	70 / 720		5630
3. POST TENSION		80 / 1280	90 / 1170	90 / 1260	90 / 360	90 / 1080	90 / 720		5870
4. SLURRY WALL		90 / 1440	90 / 1170	90 / 1260	100 / 400	90 / 1080	80 / 640	8500 LF	5990
5. NO TUNNEL		100 / 1000	100 / 1300	100 / 1400	100 / 400	100 / 1200	65 / 520		6420
6. CIRCULAR SEC. (PIPE)		80 / 1200	90 / 1170	50 / 700	40				
7. ELLIPTICAL		80	70	50	40				
8. CORRUGATED STEEL		80	70	50	40				
9. PRE-CAST CULVERTS		80	90	70	40				
10.									

Excellent - 5 Very Good - 4 Good - 3 Fair - 2 Poor - 1

Seek The Best - Not Perfection

Figure 10.14 This is the form used to evaluate ideas from the idea generation stage (Table 10.2) The weights are explicitly given in the top row, and then unweighted scores are applied to each alternative. The product of the scores and weights are summed in the right-most column. Note that most of the tunnel alternatives actually score worse than no tunnel in this evaluator's assessment.

concerns that the other alternatives might, such as functional or structural considerations. This is somewhat of a joke on the part of the engineers, however, and serves more as a "standard candle" than a meaningful suggestion of how to pass a roadway under City Square. Notwithstanding the null alternative, each of the other surviving alternatives is relatively close in scoring, which suggests that another approach to measuring tunnel performance or other selection criteria is warranted. The point is, the process is valuable, but it cannot substitute for sound judgment and experienced thought.

10.3 1998: CITY SQUARE AS ENGINEERS SEE IT NOW

Ultimately, the many studies and design exercises resulted in the adoption of a tunnel under City Square that restored that area into a vital part of the City of Boston. Removal of the elevated highway from the heart of the Charlestown community allowed that community to restore its historic relationship with the waterfront. Figure 10.15 is a sketch of the City Square area by the same artist as Figure 10.2. Note that the community is now able to develop around the Bunker Hill monument and continue to the point where the U.S.S. Constitution is berthed. It should be mentioned, however, that this view can be observed by anyone unfortunate enough to be stalled on the elevated highways that still continue to run along the southern perimeter of Charlestown. As such, the view is perhaps more idyllic than the actual place. Residents choosing to walk from City Square to the waterfront are never entirely out of view or earshot of an elevated highway.

What design decisions were made to implement the conceptual designs for the tunnel under City Square? It was actually built as a reinforced concrete box tunnel. The connection to the south was made using trumpet interchanges that both match with nearby highway designs and minimize impacts on property. The alternative for the streets that was adopted was the one favored by the residents of Charlestown and the City of Boston: It limited the number of off-ramps from the highway into City Square to interfere with the tendency of commuters to view City Square as a regular bypass into the city. The project was financed and completed as the first part of the Central Artery Project, which is still underway in Boston—and is expected to last until the year 2010!

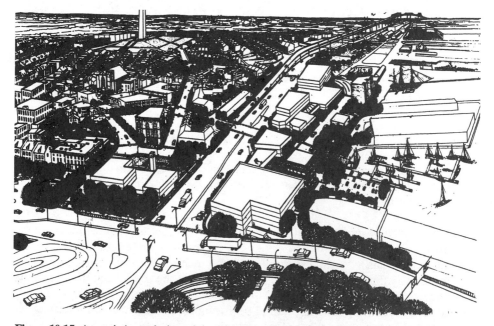

Figure 10.15 An artist's rendering of the City Square area after the project was completed. Compare this with the view by the same artist in Figure 10.2. From (Berger 1981).

After almost fifty years of discussion, debate, and design activity, City Square has been returned from a transportation-dominated eyesore to a vibrant community space that fulfills both community and transportation needs. It can be argued that the design process that was followed ultimately resulted in a "win-win" situation for all of the stakeholders.

10.4 NOTES

Section 10.1: The primary documentary sources for this chapter are (Barton-Aschman 1962), (Massachusetts Department of Public Works 1974), and (Berger 1981). In addition, a number of individuals were helpful in explaining the history and the process. Of these, the most valuable was Fred Salvucci, former Secretary of Transportation for the State of Massachusetts and now on the faculty of the Massachusetts Institute of Technology.

Section 10.1.1: Save for Figure 10.2, taken from (Berger 1981), and three objectives trees drawn by the authors (Figures 10.5, 10.9, and 10.10), the figures in this section are from (Barton-Aschman 1962).

Section 10.1.2: (Massachusetts Department of Public Works 1974) is the primary source for this section. Fred Salvucci's comments on the impact on his personal and professional life are from a private conversation.

Section 10.1.3: The title of this section comes from a statement made by State Senator McKenna at a public hearing in Charlestown, Oct. 24, 1972. After expressing opposition to a roads-based plan without what he considered sufficient community input, he said that "Charlestown is only now returning as an honorable place to live after years of struggle to improve." The primary source for this section is (Berger 1981).

Section 10.2: The examples included were used as appendices in (Berger 1981).

10.5 EXERCISES

10.1 Describe the evolution of the engineers' understanding of who were the stakeholders in the City Square transportation project.

10.2 How did the evolution described in Exercise 10.1 affect the design process?

10.3 Compare the graphic display of objectives and concerns for the City Square project (Figure 10.4) with the objectives tree of Figure 10.5. What are their respective advantages and disadvantages?

10.4 Generate an alternative approach to evaluating design ideas for tunnels and contrast your results with those given in Figure 10.14.

10.5 Find a publicly funded project in your city or town and ask the engineers on that project how they have identified the stakeholders, and determined the objectives, constraints, and functions for that project?

Bibliography

In addition to providing the references cited in the text, the bibliography below is a sampling of books covering a wide range of issues including design theory, design in different disciplines, project management techniques, optimization theory, some applications of artificial intelligence, engineering ethics and the practice of engineering, and more. This list of works is *not* complete—the literatures on design and project management alone are both vast and rapidly expanding. Thus, it should be kept in mind that this bibliography represents only the tip of a very large iceberg of published work in design and in project management. Some of the works cited are just intellectually interesting, and some are books that students in particular will find useful for project work.

J. L. Adams, *Conceptual Blockbusting: A Guide to Better Ideas*, Stanford Alumni Association, Stanford, CA, 1979.

K. Akiyama, *Function Analysis: Systematic Improvement of Quality and Performance*, Productivity Press, Cambridge, MA, 1991.

C. Alexander, *Notes on the Synthesis of Form*, Harvard University Press, Cambridge, MA, 1964.

Anon., *Goals and Priorities for Research in Engineering Design*, American Society of Mechanical Engineers, New York, NY, 1986.

Anon., *Improving Engineering Design: Designing for Competitive Advantage*, National Research Council, National Academy Press, Washington, D.C., 1991.

Anon., *Managing Projects and Programs*, The Harvard Business Review Book Series, Harvard Business School Press, Cambridge, MA, 1989.

H. Arendt, *Eichmann in Jerusalem: A Report on the Banality of Evil*, Viking Press, New York, NY, 1963.

J. S. Arora, *Introduction to Optimum Design*, McGraw-Hill, New York, NY, 1989.

W. Asimow, *Introduction to Design*, Prentice-Hall, Englewood Cliffs, NJ, 1962.

A. B. Badiru, *Project Management in Manufacturing and High Technology Operations*, John Wiley & Sons, New York, NY, 1996.

K. M. Bartol and D. C. Martin, *Management*, 2nd edit., McGraw-Hill Book Company, New York, NY, 1994.

Barton-Aschman Associates, *North Area Terminal Study*, Technical Report, Barton-Aschman Associates, Evanston, IL, August 1962.

Louis Berger, *Central Artery North Area Project*, Interim Report, Louis Berger & Associates, Cambridge, MA, 1981.

G. Boothroyd and P. Dewhurst, *Product Design for Assembly*, Boothroyd Dewhurst Inc., Wakefield, RI, 1989.

C. L. Bovee, M. J. Houston, and J. V. Thill, *Marketing*, 2nd edit., McGraw-Hill Book Company, New York, NY, 1995.

C. L. Bovee, J. V. Thill, M. B. Word, and G. P. Dovel, *Management*, McGraw-Hill Book Company, New York, NY, 1993.

E. T. Boyer, F. D. Meyers, F. M. Croft, Jr., M. J. Miller, and J. T. Demel, *Technical Graphics*, John Wiley & Sons, New York, NY, 1991.

D. C. Brown, "Design," in S. C. Shapiro (Editor), *Encyclopedia of Artificial Intelligence*, 2nd Edition, John Wiley & Sons, New York, NY, 1992.

D. C. Brown and B. Chandrasekaran, *Design Problem Solving*, Pitman, London, and Morgan Kaufmann, Los Altos, CA, 1989.

L. L. Bucciarelli, *Designing Engineers*, MIT Press, Cambridge, MA, 1994.

S. Carlson Skalak, H. Kemser, and N. Ter-Minassian, "Defining a Product Development Methodology with Concurrent Engineering for Small Manufacturing Companies," *Journal of Engineering Design*, 8 (4), 305-328, December 1997.

A. D. S. Carter, *Mechanical Reliability*, Macmillan, London, England, 1986.

J. Connor, K. Kubler, P. Leitzell, J. P. Strozzo, and M. Wang, *Design of a Chicken Coop*, E4 Project Report, Department of Engineering, Harvey Mudd College, Claremont, CA, 1997.

J. Corbett, M. Dooner, J. Meleka, and C. Pym, *Design for Manufacture: Strategies, Principles and Techniques*, Addison-Wesley, Wokingham, England, 1991.

R. D. Coyne, M. A. Rosenman, A. D. Radford, M. Balachandran and J. S. Gero, *Knowledge-Based Design Systems*, Addison-Wesley, Reading, MA, 1990.

N. Cross, *Engineering Design Methods*, 2nd edit., John Wiley, Chichester, England, 1994.

M. L. Dertouzos, R. K. Lester, R. M. Solow and the MIT Commission on Industrial Productivity, *The Making of America: Regaining the Productive Edge*, MIT Press, Cambridge, MA, 1989.

J. R. Dixon, *Design Engineering: Inventiveness, Analysis, and Decision Making*, McGraw-Hill, New York, NY, 1966.

J. R. Dixon, "Engineering Design Science: The State of Education," *Mechanical Engineering*, 113 (2), February 1991.

J. R. Dixon, "Engineering Design Science: New Goals for Education," *Mechanical Engineering*, 113 (3), March 1991.

J. R. Dixon and C. Poli, *Engineering Design and Design for Manufacturing*, Field Stone Publishers, Conway, MA, 1995.

C. L. Dym (Editor), *Applications of Knowledge-Based Systems to Engineering Analysis and Design*, American Society of Mechanical Engineers, New York, NY, 1985.

C. L. Dym (Editor), *Computing Futures in Engineering Design*, Harvey Mudd College, Claremont, CA, 1997.

C. L. Dym, *E4 (Engineering Projects) Handbook*, Department of Engineering, Harvey Mudd College, Claremont, CA, Spring 1993.

C. L. Dym, *Engineering Design: A Synthesis of Views*, Cambridge University Press, New York, NY, 1994a.

C. L. Dym, Letter to the Editor, *Mechanical Engineering*, 114 (8), August 1992.

C. L. Dym, "The Role of Symbolic Representation in Engineering Education," *IEEE Transactions on Education*, 35 (2), March 1993.

C. L. Dym, "Teaching Design to Freshmen: Style and Content," *Journal of Engineering Education*, 83 (4), 303–310, October 1994b.

C. L. Dym and E. S. Ivey, *Principles of Mathematical Modeling*, Academic Press, New York, NY, 1980.

C. L. Dym and R. E. Levitt, *Knowledge-Based Systems in Engineering*, McGraw-Hill, New York, NY, 1991.

C. E. Ebeling, *An Introduction to Reliability and Maintainability Engineering*, McGraw-Hill, New York, NY, 1997.

D. L. Edel, Jr., (Editor), *Introduction to Creative Design*, Prentice-Hall, Englewood Cliffs, NJ, 1967.

K. S. Edwards, Jr., and R. B. McKee, *Fundamentals of Mechanical Component Design*, McGraw-Hill, New York, NY, 1991.

K. A. Ericsson and H. A. Simon, *Protocol Analysis: Verbal Reports as Data*, MIT Press, Cambridge, MA, 1984.

A. Ertas and J. C. Jones, *The Engineering Design Process*, John Wiley & Sons, New York, NY, 1993.

D. L. Evans (Coordinator), "Special Issue: Integrating Design Throughout the Curriculum," *Engineering Education*, *80* (5), 1990.

J. H. Faupel, *Engineering Design*, John Wiley & Sons, New York, NY, 1964.

R. L. Fox, *Optimization Methods for Engineering Design*, Addison-Wesley, Reading, MA, 1971.

J. Fortune and G. Peters, *Learning From Failure-The Systems Approach*, John Wiley, Chichester, England, 1995.

M. E. French, *Conceptual Design for Engineers*, 2nd Edition, Design Council Books, London, England, 1985.

M. E. French, *Form, Structure and Mechanism*, MacMillan, London, England, 1992.

D. C. Gause and G. M. Weinberg, *Exploring Requirements: Quality Before Design,* Dorset House Publishing, New York, NY, 1989.

J. S. Gero (Editor), *Design Optimization*, Academic Press, Orlando, FL, 1985.

J. S. Gero (Editor), *Proceedings of AI in Design '92*, Kluwer Academic Publishers, Dordrecht, The Netherlands, 1992.

J. S. Gero (Editor), *Proceedings of AI in Design '94*, Kluwer Academic Publishers, Dordrecht, The Netherlands, 1994.

J. S. Gero (Editor), *Proceedings of AI in Design '96*, Kluwer Academic Publishers, Dordrecht, The Netherlands, 1996.

M. P. Glazer and P. M. Glazer, *The Whistleblowers: Exposing Corruption in Government and Industry*, Basic Books, New York, NY, 1989.

G. L. Glegg, *The Design of Design*, Cambridge University Press, Cambridge, England, 1969.

G. L. Glegg, *The Science of Design*, Cambridge University Press, Cambridge, England, 1973.

G. L. Glegg, *The Selection of Design*, Cambridge University Press, Cambridge, England, 1972.

T. J. Glover, *Pocket Ref*, Sequoia Publishing, Littleton, CO, 1993.

S. H. Goldstein and R. A. Rubin, "Engineering Ethics," *Civil Engineering*, October 1996.

P. Graham (Editor), *Mary Parker Follett—Prophet of Management: A Celebration of Writings From the 1920s*, Harvard Business School Press, Boston, MA, 1996.

P. Gutierrez, J. Kimball, B. Maul, A. Thurston, and J. Walker, *Design of a Chicken Coop*, E4 Project Report, Department of Engineering, Harvey Mudd College, Claremont, CA, 1997.

C. Hales, *Managing Engineering Design*, Longman Scientific & Technical, Harlow, England, 1993.

J. Harr, *A Civil Action*, Vinatge Books, New York, NY, 1995.

B. Hartmann, B. Hulse, S. Jayaweera, A. Lamb, B. Massey, and R. Minneman, *Design of a "Building Block" Analog Computer*, E4 Project Report, Department of Engineering, Harvey Mudd College, Claremont, CA, 1993.

S. I. Hayakawa, *Language in Thought and Action*, 4th Edition, Harcourt Brace Jovanovich, San Diego, CA, 1978.

R. T. Hays, "Value Management," in W. K. Hodson (Editor), *Maynard's Industrial Engineering Handbook*, 4th Edition, McGraw-Hill Book Company, New York, NY, 1992.

J. Heskett, *Industrial Design*, Thames and Hudson, London, 1980.

R. S. House, *The Human Side of Project Management*, Addison-Wesley, Reading, MA, 1988.

V. Hubke, M. M. Andreasen, and W. E. Eder, *Practical Studies in Systematic Design*, Butterworths, London, England, 1988.

B. Hyman, *Topics in Engineering Design*, Prentice Hall, Englewood Cliffs, NJ, 1998.

D. Jain, G. P. Luth, H. Krawinkler, and K. H. Law, *A Formal Approach to Automating Conceptual Structural Design*, Technical Report No. 31, Center for Integrated Facility Engineering, Stanford University, Stanford, CA, 1990.

F. D. Jones, *Ingenious Mechanisms*: Vols. 1–3, The Industrial Press, New York, NY, 1930.

J. C. Jones, *Design Methods*, Wiley-Interscience, Chichester, England, 1992.

D. Kaminski, "A Method to Avoid the Madness," *The New York Times*, 3 November 1996.

H. Kerzner, *Project Management: A Systems Approach to Planning, Scheduling and Controlling*, Van Nostrand Reinhold, New York, NY, 1992.

D. S. Kezsbom, D. L. Schilling, and K. A. Edward, *Dynamic Project Management: A Practical Guide for Managers & Scientists*, John Wiley & Sons, New York, NY, 1989.

M. Levy and M. Salvadori, *Why Buildings Fall Down*, Norton, New York, NY, 1992.

E. E. Lewis, *Introduction to Reliability Engineering*, John Wiley & Sons, New York, NY, 1987.

P. Little, *Improving Railroad Car Reliability Using a New Opportunistic Maintenance Heuristic and Other Information System Improvements*, Doctoral Dissertation, Massachusetts Institute of Technology, Cambridge, MA, 1991.

M. W. Martin and R. Schinzinger, *Ethics in Engineering*, 3rd Edition, McGraw-Hill Book Company, New York, NY, 1996.

Massachusetts Department of Public Works, *North Terminal*, Draft Environmental Impact Report (Section 4(F) and Section 106 Statements), Massachusetts Department of Public Works, Boston, MA, 1974.

R. L. Meehan, *Getting Sued and Other Tales of the Engineering Life*, The MIT Press, Cambridge, MA, 1981.

J. R. Meredith and S. J. Mantel, Jr., *Project Management: A Managerial Approach*, John Wiley & Sons, New York, NY, 1995.

J. Morgenstern, "The Fifty-nine-Story Crisis," *The New Yorker*, 29 May 1995.

T. T. Nagle, *The Strategy and Tactics of Pricing*, Prentice-Hall, Englewood Cliffs, NJ, 1987.

A. Newell and H. A. Simon, *Human Problem Solving*, Prentice-Hall, Englewood Cliffs, NJ, 1972.

G. D. Oberlander, *Project Management for Engineering and Construction*, McGraw-Hill, New York, NY, 1993.

G. Pahl and W. Beitz, *Engineering Design: A Systematic Approach*, 2nd Edition, Springer, London, England, 1996.

A. Palladio, *The Four Books of Architecture*, Dover, New York, NY, 1965.

Y. C. Pao, *Elements of Computer-Aided Design and Manufacturing*, John Wiley & Sons, New York, NY, 1984.

P. Y. Papalambros and D. J. Wilde, *Principles of Optimal Design: Modeling and Computation*, Cambridge University Press, Cambridge, England, 1988.

H. Petroski, *Design Paradigms*, Cambridge University Press, New York, NY, 1994.

H. Petroski, *Engineers of Dreams*, Alfred A. Knopf, New York, NY, 1995.

H. Petroski, *To Engineer Is Human*, St. Martin's Press, New York, NY, 1985.

L. Phlips, *The Economics of Price Discrimination*, Cambridge University Press, Cambridge, England, 1985.

S. Pugh, *Total Design: Integrated Methods for Successful Product Engineering*, Addison-Wesley, Wokingham, England, 1991.

H. E. Riggs, *Financial and Cost Analysis for Engineering and Technology Management*, John Wiley & Sons, New York, NY, 1994.

J. L. Riggs and T. M. West, *Essentials of Engineering Economics*, McGraw-Hill, New York, NY, 1986.

M. D. Rychener (Editor), *Expert Systems for Engineering Design*, Academic Press, Boston, MA, 1988.

M. Salvadori, *Why Buildings Stand Up*, McGraw-Hill, New York, NY, 1980.

D. A. Schon, *The Reflective Practitioner*, Basic Books, New York, NY, 1983.

D. Schroeder, "Little Land Bruisers," *Car and Driver*, 96–109, May 1998.

R. G. Schroeder, *Operations Management: Decision Making in the Operations Function*, McGraw-Hill, New York, NY, 1993

J. J. Shah, "Experimental Investigation of Progressive Idea Generation Techniques in Engineering Design," *Proceedings of the 1998 ASME Design Theory and Methodology Conference*, American Society of Mechanical Engineers, New York, NY, 1998.

H. A. Simon, "Style in Design," in C. M. Eastman (Editor), *Spatial Synthesis in Computer-Aided Building Design*, Applied Science Publishers, London, England, 1975.

H. A. Simon, *The Sciences of the Artificial*, 2nd Edition, MIT Press, Cambridge, MA, 1981.

L. Stauffer, *An Empirical Study on the Process of Mechanical Design*, Thesis, Department of Mechanical Engineering, Oregon State University, Corvallis, OR, 1987.

L. Stauffer, D. G. Ullman, and T. G. Dietterich, "Protocol Analysis of Mechanical Engineering Design," In *Proceedings of the 1987 International Conference on Engineering Design*. Boston, MA, 1987.

G. Stevens, *The Reasoning Architect: Mathematics and Science in Design*, McGraw-Hill, New York, NY, 1990.

G. Stiny and J. Gips, *Algorithmic Aesthetics*, University of California Press, Berkeley, CA, 1978.

N. P. Suh, *The Principles of Design*, Oxford University Press, Oxford, England, 1990.

M. C. Thomsett, *The Little Black Book of Project Management*, American Management Association, New York, NY, 1990.

C. Tong and D. Sriram (Editors), *Artificial Intelligence in Engineering Design, Volume I: Design Representation and Models of Routine Design*, Academic Press, Boston, MA, 1992a.

C. Tong and D. Sriram (Editors), *Artificial Intelligence in Engineering Design, Volume II: Models of Innovative Design, Reasoning about Physical Systems, and Reasoning about Geometry*, Academic Press, Boston, MA, 1992b.

C. Tong and D. Sriram (Editors), *Artificial Intelligence in Engineering Design, Volume III: Knowledge Acquisition, Commercial Applications and Integrated Environments*, Academic Press, Boston, MA, 1992c.

B. W. Tuckman, "Developmental Sequences in Small Groups," *Psychological Bulletin*, *63*. 384–399. 1965.

D. G. Ullman, "A Taxonomy for Mechanical Design," *Research in Engineering Design*, *3*. 1992.

D. G. Ullman, *The Mechanical Design Process*, 2nd Edition, McGraw-Hill, New York, NY, 1997.

D. G. Ullman and T. G. Dietterich, "Toward Expert CAD," *Computers in Mechanical Engineering*, *6* (3), 1987.

D. G. Ullman, T. G. Dietterich, and L. Stauffer, "A Model of the Mechanical Design Process Based on Empirical Data," *Artificial Intelligence for Engineering Design, Analysis and Manufacturing*, *2* (1), 1988.

D. G. Ullman, S. Wood, and D. Craig, "The Importance of Drawing in the Mechanical Design Process," *Computers and Graphics*, *14* (2), 1990.

K. T. Ulrich and S. D. Eppinger, *Product Design and Development*, McGraw-Hill, New York, NY, 1995.

G. N. Vanderplaats, *Numerical Optimization Techniques for Engineering Design*, McGraw-Hill, New York, NY, 1984.

VDI, *VDI-2221: Systematic Approach to the Design of Technical Systems and Products*, Verein Deutscher Ingenieure, VDI-Verlag, Translation of the German Edition 11/1986, 1987.

C. E. Wales, R. A. Stager, and T. R. Long, *Guided Engineering Design: Project Book*, West Publishing Company, St. Paul, MN, 1974.

J. Walton, *Engineering Design: From Art to Practice*, West Publishing, St. Paul, MN, 1991.

D. J. Wilde, *Globally Optimal Design*, John Wiley & Sons, New York, NY, 1978.

T. T. Woodson, *Introduction to Engineering Design*, McGraw-Hill, New York, NY, 1966.

R. N. Wright, S. J. Fenves, and J. R. Harris, *Modeling of Standards: Technical Aids for Their Formulation, Expression and Use*, National Bureau of Standards, Washington, D.C., March 1980.

C. Zener, *Engineering Design by Geometric Programming*, Wiley-Interscience, New York, NY, 1971.

C. Zozaya-Gorostiza, C. Hendrickson, and D. R. Rehak, *Knowledge-Based Process Planning for Construction and Manufacturing*, Academic Press, Boston, MA, 1989.

Photo Credits

Chapter 1

Figure 1.2a: Courtesy Lockheed Advanced Development Company. Figure 1.2b: Courtesy British Airways. Figure 1.2c: Courtesy The Boeing Corporation. Figure 1.2d: Courtesy AeroVironment, Inc. Figure 1.3a: John Welzenbach/The Stock Market. Figure 1.3b: Phyllis Picardi/Stock Boston/Picturequest. Figure 1.3c: Jose Carrillo/PhotoEdit/Picturequest. Figure 1.3d: Alan Leveson/Tony Stone Images. Figure 1.4a: Pascal Crapet/Tony Stone Images. Figure 1.4b: G. K. & Vikki Hart/The Image Bank. Figure 1.4c: Terje Rakke/The Image Bank. Figure 1.4d: ©Yeager/The Stock Market.

Chapter 6

Figure 6.1a: Courtesy Johnson & Johnson. Figure 6.10: Courtesy Clive Dym.

Chapter 7

Figure 7.5 and Figure 7.6: Courtesy Clive Dym.

Chapter 9

Figure 9.3 and Figure 9.4: Courtesy Clive Dym.

Index

A

Activities
  critical path, 98
  slack, 98
Activity networks, 86, 96–100
Activity-on-node (AON), 96–100
Adjoining (adjacent) ideas, 144–145
Affordability, design for, 204–210
  cost estimation, 205, 208–209
  costing and pricing, 210
  time value of money, 205–207
Agency-loyalty, 223–224
American Society of Civil Engineers
  (ASCE), 220–221, 223, 224
Amortization, of costs, 210
Analogies between situations, 42,
  143–144
Analysis of design, 24, 35
  cost analysis, 207
  methods of, 43–44
Architecture of design, 34
Artifacts, 8
  attributes of (*See* Attributes)
  boundaries of system, 115
  choosing type of, 27
  form of, 8, 10, 13–15, 26
  functions of (*See* Functions)
  goals, 57 (*See also* Objectives)
  maintainability, 214–215
  means or implementations
    (*See* Means or implementations)

  pricing of, 210
  reliability/failure of, 211–214
  specifications (*See* Specifications)
Assembly, design for, 17, 202–203
Assembly chart, 204
Assembly drawings, 186, 188
Attributes, 8–9, 55–57, 58
  characterized by objectives, 119
  "design for X (the -ilities)", 200–201
  freezing number of, 146
  list, 56–57, 63–64
  requirements matrix, 39
  solution-independent, 39
Availability of different technologies, 145

B

Benchmarking, 42–43, 139
Best of class chart, 150, 154
Beta testing, 44
Beverage container project, xiii, 3, 8, 20,
  52, 54–55
  activity-on-node network, 99
  attributes list, 63–64
  functions, 123–124
  Gantt charts, 101
  linear responsibility chart (LRC), 93
  management of, 84
  morphological chart, 147, 148
  numerical evaluation matrices,
    150–152
  objectives tree, 62–64, 72–73